Edward Kostiner, Ph.D.
Former Professor of Chemistry
University of Connecticut
Storrs, Connecticut

Neil Jespersen, Ph.D.
Professor of Chemistry
St. John's University
Jamaica, New York

Chemistry

Second Edition

All inquiries should be addressed to:
Barron's Educational Series, Inc.
250 Wireless Boulevard
Hauppauge, New York 11788
http://www.barronseduc.com

Library of Congress Catalog Card No. 2002027995

International Standard Book No. 0-7641-2006-9

Library of Congress Cataloging-in-Publication Data
Kostiner, Edward.
    Chemistry / Edward Kostiner, Neil Jespersen.—2nd ed.
      p.   cm.—(Barron's EZ-101 study keys)
    Includes index.
    ISBN 0-7641-2006-9 (alk. paper)
    1. Chemistry—Study and teaching (Higher)
I. Jespersen, Neil D. II. Title. III. Barron's EZ 101
study keys.
QD40 .K67   2003
540—dc21   2002027995

PRINTED IN THE UNITED STATES OF AMERICA
9 8 7 6 5 4 3 2

CONTENTS

Theme 1 THE SCIENCE
OF CHEMISTRY

*C*hemistry is the study of matter (anything that has mass and occupies space) and the interactions among different types of matter. The goal of the chemist is to understand what matter is made of and how it is organized and structured.

In everyday life, you observe **macroscopic** (or bulk) properties of substances—those that can be easily measured or observed. Chemistry attempts to explain macroscopic properties on the basis of the **microscopic** behavior of atoms and molecules—the basic building blocks of chemical matter.

INDIVIDUAL KEYS IN THIS THEME

1	Fields of chemistry
2	The scientific method

Key 1 Fields of chemistry

OVERVIEW *Historically, the field of chemistry was considered to be divided into four major areas. However, with continued progress, many areas of specialization have been defined.*

The four classical areas of chemistry are

- **Organic chemistry:** The chemistry of the element carbon and its compounds (the term *organic* comes from the Greek word for living matter).
- **Inorganic chemistry:** The chemistry of all the other elements (*inorganic* means inanimate or nonliving).
- **Analytical chemistry:** The qualitative (What is it?) and the quantitative (How much is there?) aspects of chemistry—the chemical analysis of compounds.
- **Physical chemistry:** The quantitative aspects of the physical properties of substances and their relationships to chemical structure and composition.

More recently, other areas of specialization have been defined, for example,

- **Biochemistry:** The chemistry of biologically active substances.
- **Polymer chemistry:** The chemistry of polymeric (long-chain) molecules.
- **Solid state chemistry:** The chemical behavior of solids.
- **Organometallic chemistry:** The chemistry of compounds in which a metal is bonded in an organic compound.

Key 2   The scientific method

OVERVIEW *The scientific method is a logical iterative process by which progress in any science is made.*

The stepwise nature of the scientific method is illustrated as follows:

1. **Experiments** are carried out to determine the nature of a substance or process. Careful observations, either qualitative (general observations) or quantitative (involving measurements), are made and form the basis for scientific investigations.
2. **Hypotheses** (or tentative explanations) are formulated to describe these experimental observations.
3. **Scientific laws** are developed to concisely summarize (usually in mathematical form) relationships developed after collecting a large amount of experimental data.
4. **Scientific theories** are developed to provide unifying concepts among several laws. Scientific theories must be testable and disprovable. Often the theory itself suggests the existence of new laws and consequently experiments that will test the validity of the theory.

Once a theory is developed, more experiments are designed to ensure its validity. If a theory is found to be inadequate, it is modified or even discarded. Science progresses in this iterative fashion, a series of steps continuously modifying current theories to reflect new and perhaps contradictory experimental evidence.

Theme 2 MEASUREMENTS

*C*hemistry is a quantitative science. Chemists must be capable of making precise and reproducible measurement with devices or instruments designed for a particular task. Measurements should be reported in a common system of units and should reflect the quality of the data. Finally, problem solving is greatly facilitated by the use of dimensional (or factor label) analysis.

Key 3 The International System of Units

OVERVIEW *The International System of Units (or SI) provides a precisely defined set of seven base units from which all other units can be derived. Metric prefixes allow very large and very small values to be written conveniently.*

A metric prefix is a single letter (or symbol) preceding the abbreviated SI unit it modifies. For quick mathematical conversions, the metric prefix and its exponential value are interchangeable.

Metric Prefixes

Name	Prefix	Value
giga	G	$\times 10^9$
mega	M	$\times 10^6$
kilo	k	$\times 10^3$
deci	d	$\times 10^{-1}$
centi	c	$\times 10^{-2}$
milli	m	$\times 10^{-3}$
micro	μ	$\times 10^{-6}$
nano	n	$\times 10^{-9}$
pico	p	$\times 10^{-12}$

Key 4   Length and mass

OVERVIEW *The SI units of length and mass and units derived from them will be considered along with devices for their measurement.*

Length has the **meter (m)** as its base unit. This unit is usually too large for laboratory work, where the **centimeter** ($1 \text{ cm} = 1 \times 10^{-2}$ meter) and the **millimeter** ($1 \text{ mm} = 1 \times 10^{-3}$ meter) are used. On the atomic scale, chemists use the **picometer** ($1 \text{ pm} = 1 \times 10^{-12}$ meter) and the **nanometer** ($1 \text{ nm} = 1 \times 10^{-9}$ meter).

Length is usually measured with a meterstick, a common 12-inch ruler, or more accurate calipers. Relating the SI to the English system, one inch is exactly 2.54 cm and one meter is approximately 39 inches or 1.1 yards.

Volume is always a **derived unit**. The **cubic meter (m^3)** is too large for convenient use, and the **cubic decimeter (dm^3)** is the preferred unit. The dm^3 is equal in volume to the traditional **liter (L)**. By definition, 10 centimeters is equal to one decimeter. As a result, $1 \text{ L} = 1000 \text{ mL} = 1 \text{ dm}^3 = 1000 \text{ cm}^3$ (10 cm × 10 cm × 10 cm).

Volume is usually measured with

- **graduated cylinders**, used to measure approximate volumes
- **pipets**, calibrated for exact delivery of liquids
- **burets**, graduated tubes with a valve to dispense and measure exact volumes
- **volumetric flasks**, used to contain exact volumes of solution.

Relating the units of volume to English units shows that one liter is a little more than one quart and that 250 mL is approximately one cup.

Mass has the SI unit of the **kilogram**, which is also too large for laboratory work. The common units used in the laboratory are the **gram** and the **milligram** ($1 \text{ mg} = 1 \times 10^{-3} \text{ g}$).

Weight and mass are not the same thing. **Mass** is the amount of matter in a given sample, while **weight** is the force of gravity acting on the sample. Mass is determined by weighing a sample on a balance calibrated with known masses. Weight is the result of the force of gravity acting on a mass (weight = force × mass). On a two-pan balance, known masses are balanced with the sample so that

$$(\text{force}_{\text{gravity}})(\text{mass}_{\text{known}}) = (\text{force}_{\text{gravity}})(\text{mass}_{\text{sample}})$$

The two sides are balanced by adjusting the amount of known mass. The force of gravity cancels, and the result is that a weighing process determines the mass of the sample.

Density is a physical property often used to identify a substance. It is the ratio of the mass to the volume of any material (D = mass/volume). The common units of density are g/cm^3 for solids and liquids and g/L for gases.

Specific gravity is the ratio of the density of a substance in g/cm^3 divided by the density of water at 25°C, which is $1.00\,g/cm^3$. The result is a **dimensionless** unit.

Key 5 Pressure and temperature

OVERVIEW *Pressure is one of the common quantities measured in the laboratory. It has particular importance when considering the behavior of matter in the gas phase. There are three temperature scales in common use. Scientists normally use either the Celsius (centigrade) scale or the absolute temperature scale.*

Pressure is defined as the force (mass × acceleration) per unit area.

The **pascal (Pa)** is the unit of pressure in the SI. $1\,Pa = 1\,N/m^2$, where N is the **newton**, the derived unit of force that has SI units of $kg\,m/s^2$.

More commonly used units of pressure are

- one **bar** equals $10^5\,Pa$.
- **atmospheric pressure** is the pressure exerted by the atmosphere and equals the pressure exerted by a column of mercury (Hg) 760 mm high.
- one **torr** is the pressure exerted by a 1-mm column of Hg; 760 torr = 1 atm.

A **manometer** is an instrument used to measure pressure.

A **barometer** is an instrument used to measure atmospheric pressure.

As a body gains heat (or thermal energy), it goes to a state of higher **temperature**, which is just a measure of how hot an object is.

Three temperature scales are in common use:

- The **Fahrenheit scale (°F)** is the scale on which 32°F is defined by the freezing point of water and 212°C is defined by the boiling point of water.
- The **Celsius scale (°C)** is the scale on which 0°C is defined by the freezing point of water and 100°C is defined by the boiling point of water.

To convert from Celsius to Fahrenheit, use the following formula:

$$°F = \frac{9}{5}°C + 32$$

To convert from Fahrenheit to Celsius, use the following formula:

$$°C = \frac{5}{9}(°F - 32)$$

On the Kelvin or absolute temperature scale, one degree has the same magnitude as one degree on the Celsius scale. However, the zero point of the Kelvin scale is absolute zero or −273.16°C.

Absolute zero: 0.0 K (−273.16°C) is the lowest temperature attainable. A Kelvin temperature can never be a negative value.

Key 6 Significant figures

OVERVIEW *In an experimental science, it is necessary to distinguish between the precision and the accuracy of a measurement and to be able to report the result of calculations to reflect actual experimental conditions.*

The **accuracy** of a measurement tells you how close the measurement is to the true (or accepted true) value. The difference between the true value and the measurement is called the **error**.

The **precision** of a measurement evaluates how close repeated measurements are to each other. The difference between the average of a series of measurements and an individual measurement is known as the **deviation**.

The number of **significant figures** in a reported quantity or number is the number of meaningful digits in that measured quantity. The precision of a measurement is reflected by the number of significant figures reported.

Measurements always involve an estimation of the last digit in the number and have an **uncertainty** of ± 1 in the last digit of the reading. Some numbers have no uncertainty and are called **exact numbers**. Exact numbers are often conversion factors (e.g., there are exactly 12 inches in one foot).

The rules for determining which digits of a number are significant are

1. Digits other than zero are **always significant**.
2. Zeros between two nonzero digits are **always significant**.
3. Zeros to the left of all nonzero digits are **never significant**.
4. Zeros to the right of the last nonzero digit are **always significant** if the number contains a decimal point.
5. Zeros to the right of the last nonzero digit **may be significant** if the number does not contain a decimal point. This ambiguity can be eliminated by using scientific notation. For example, 320,000 has at least two significant figures. Writing the number as 3.20×10^5 gives three significant figures, 3.200×10^5 gives four significant figures, 3.2000×10^5 gives five significant figures, and 3.20000×10^5 gives six significant figures.

Key 7 Conversion factors and

dimensional analysis

OVERVIEW *It is of particular importance to be able to convert from one unit to another. Each of the relations between units given in Keys 3, 4, and 5, for example, is a conversion factor.*

A **conversion factor** is obtained from a predefined equality such as 12 inches = 1 foot or 60 seconds = 1 minute. These equalities can be made into ratios by dividing by one side or the other to give a ratio equal to 1. For example,

$$12 \text{ inches} = 1 \text{ foot}$$

$$\frac{12 \text{ inches}}{1 \text{ foot}} = 1 \qquad 1 = \frac{1 \text{ foot}}{12 \text{ inches}}$$

Either ratio can be used as a conversion factor as long as the units cancel properly. To determine the number of inches in 13.0 feet, we calculate

$$13.0 \text{ feet} \frac{12 \text{ inches}}{1 \text{ foot}} = 156 \text{ inches}$$

Some experimental properties of matter such as the density (g/cm^3) and molarity (mol solute/L solution) can be used as conversion factors. Density allows us to convert back and forth between mass and volume. Molarity gives us a way to convert between volume of solution and moles of solute.

Dimensional (or factor-label) analysis is the use of appropriate conversion factors to solve problems. By using dimensional analysis (i.e., utilizing all conversion factors in finding the solution to a numerical problem), one is able to check whether the answer is in the proper form before carrying out the calculation.

KEY PROBLEM

As an example of the use of density as a conversion factor, calculate the mass of a block of aluminum (Al) that has a volume of $12.4 \, cm^3$, given that the density of Al is $2.70 \, g/cm^3$.

$$\text{density} : \frac{g}{cm^3} = \qquad 2.70 = \frac{g}{12.4}$$

11

The density of Al can be used to directly calculate the mass from the volume of this object:

$$12.4 \text{ cm}^3 \times \frac{2.70 \text{ g}}{1 \text{ cm}^3} = 33.5 \text{ g}$$

Theme 3 MATTER

*M*atter is material that has mass and occupies space. All matter can be classified as being either a pure substance or a mixture. Furthermore, matter can exist as a solid, a liquid, or a gas and can undergo transition among these three phases. Chemistry can be considered to be the study of the properties of matter.

Key 8 Pure substances

OVERVIEW *A pure substance has a unique set of physical and chemical properties. These properties characterize that pure substance.*

Elements are pure substances that cannot be chemically separated or decomposed into simpler substances. There are a total of 113 known elements.

An **atom** is the smallest unit of an element having all the characteristics of that element.

Compounds, chemical combinations of elements, are pure substances in which the component elements are present in fixed proportions. They are uniquely characterized by their properties, which are different from those of the component elements.

A **molecule** is the smallest characteristic unit of a compound.

As an example, carbon tetrachloride (CCl_4) is a compound that is a dense liquid at room temperature. It can be chemically broken down into its component elements, carbon (C, a black solid) and chlorine (Cl_2, a yellow-green gas).

Chlorine is an example of an element that exists as a diatomic molecule in its elemental form (Cl_2). A **diatomic molecule** is a molecule that contains two atoms.

Key 9 Mixtures

OVERVIEW *Mixtures have variable physical properties and can consist of elements or compounds (or both) brought together in any proportion. Mixtures can be formed between any combination of gas, liquid, and solid.*

Mixtures are commonly divided into two kinds:

- **Heterogeneous** mixtures are made up of two or more physically discernible parts or phases.
- **Homogeneous** mixtures consist of components that cannot be individually detected.

Homogeneous mixtures are usually called **solutions**.

In a two-component system, the major component is called the **solvent** and the minor component is called the **solute**.

Aqueous solutions have water as the solvent even when water is the minor component.

A **colloid** (or **colloidal suspension**) is a heterogeneous mixture in which the particles of the lesser phase are intermediate in size between those in homogeneous mixtures and those in heterogeneous mixtures; typically they are on the order of 1 to 1000 nm in diameter. Unlike other heterogeneous mixtures, colloidal suspensions do not readily separate.

The **Tyndall effect** is the scattering of light by a colloidal suspension because the size of the suspended particles is similar to the wavelengths of visible light from 400 to 700 nm.

Key 10 Chemical and physical changes

OVERVIEW *Chemistry is the study of the chemical and physical properties of matter and how chemical substances undergo change or **reaction**. **Phase transitions** occur when a substance is transformed from one state (phase) to another; they are physical changes.*

Changes are classified as being either physical changes or chemical changes:

- In a **physical** change, a **physical property** (a property that can be studied without changing the identity of the substance) is being modified. Melting and boiling are examples of physical changes, and the melting and boiling points of a substance are physical properties of that substance.
- In a **chemical** change, a **chemical property** (a property that can be studied only by changing the identity of the substance) is observed. The fact that sulfur burns in oxygen to form sulfur dioxide is an example of a chemical property of sulfur. Once the reaction has taken place, the original sulfur is no longer present, having been converted to SO_2.

Properties can be classified as being intensive or extensive:

- An **intensive property** does not depend on the amount of material present, for example, the boiling point.
- An **extensive property** depends on the amount of material present; for example, mass or volume.

Elements or compounds can exist in any of the three **states of matter**: solid, liquid, or gas. For example, the compound represented by the chemical formula H_2O exists as ice (solid), water (liquid), and steam (gas).

The **melting point** is the temperature at which a solid changes to a liquid.

The **boiling point** is the temperature at which a liquid changes to a gas.

The **condensation point** is the temperature at which a gas changes to a liquid (the same temperature as the boiling point).

The **freezing point** is the temperature at which a liquid changes to a solid (the same temperature as the melting point).

The transition from a solid directly into a gas is known as **sublimation**, which occurs at the **sublimation point**.

Theme 4 ENERGY

*C*hemical or physical processes are usually accompanied by changes in energy, normally in the form of heat (thermal energy). In addition to heat, energy can take many forms that can be interconverted with each other. Visible light (a portion of the electromagnetic spectrum) can be considered both to have wavelike character and to act as a stream of massless particles.

INDIVIDUAL KEYS IN THIS THEME

Key 11 Energy

OVERVIEW *When a physical or chemical change occurs, heat (thermal energy) can be either absorbed or liberated by the system. Although energy is always conserved, it can be transferred from one object to another. The conservation laws express this in a formal manner.*

In an **exothermic** process, heat is released to the surroundings (*exo*, like *exit*, means that energy exits the system).

In an **endothermic** process, heat is absorbed from the surroundings (*endo*, like *into*, means that energy comes into the system).

The **joule (J)** is the fundamental unit of energy for chemical reactions. It corresponds to the amount of energy needed to raise one gram of water from 14.5°C to 15.5°C. The **calorie (cal)** is now considered obsolete. To convert old data, use 1 cal = 4.184 J.

The law of conservation of energy states that, in a chemical reaction, energy is always conserved and can be neither created nor destroyed.

Key 12   Kinetic and potential energy

OVERVIEW *The energy of a system can be considered as a measure of its capacity to do work. Energy can be categorized as being either kinetic energy or potential energy.*

Kinetic energy (KE) is energy associated with motion and is calculated from the formula $KE = \frac{1}{2}mv^2$, where m is the mass in kilograms and v is the velocity in meters per second. Using the stated units for mass and velocity, the combined units of $\mathbf{kg\,m^2\,s^{-2}}$ are the units for the **joule (J)**.

Potential energy (PE or V) is stored energy or energy resulting from relative position or structure. As an example, water stored behind a dam has greater potential energy than water in the ocean.

The absolute KE or PE of any sample of matter is impossible to determine. For the KE, we do not know the actual velocity of matter in our universe, and for the PE we cannot determine all the attractive and repulsive forces throughout the universe that may be acting on our sample. Changes in KE and PE (ΔKE and ΔPE) are relatively easy to measure.

The average KE is related to the absolute temperature of the sample, and changes in temperature are easy to measure. The **heat energy (q)** needed to cause a temperature change is ΔKE.

The change in potential energy can be reduced to changes in volume in gaseous reactions at constant pressure ($P\Delta V$), and these are also easy to measure. The term $P\Delta V$ is also known as the energy due to **work (w)**.

While energy can be neither created nor destroyed (Key 11), one form of energy can be **converted** to another form. As a simple example, ocean water absorbs the sun's radiant energy (Key 13), warms up (thermal energy), and evaporates into the atmosphere (potential energy). When the water condenses to rain (releasing thermal energy), falls onto a mountain, and flows into a river (potential energy to kinetic energy), it can be made to drive a turbine (kinetic energy to work) that drives a generator (mechanical energy to electrical energy).

Key 13   Electromagnetic radiation
and light

OVERVIEW *During the seventeenth century, Isaac Newton performed experiments that proved that sunlight can be split up into its component colors by passage through a prism. This occurs because the beams of each of the colors that make up sunlight (or white light) are bent to different extents as they pass through the prism. This effectively separates the beams and makes them individually visible (as in a rainbow).*

Electromagnetic radiation makes up the electromagnetic spectrum and can be considered to be composed of waves that travel at the speed of light ($c = 3.00 \times 10^8$ m s^{-1}).

Electromagnetic waves are characterized by the following:

- **wavelength** (λ, lambda), the distance between the crests of a wave.
- **frequency** (ν, nu), in s^{-1} (waves per second or reciprocal second, 1/s), the number of times the wave moves past a given reference point.

The relationship between wavelength and frequency is given by

$$\lambda \, (\text{m}) \times \nu \, (\text{s}^{-1}) = c \, (\text{m s}^{-1})$$

- **amplitude**, magnitude of the wave, corresponds to the height of the wave.

The **electromagnetic spectrum** covers a wide range of wavelengths and frequencies, from the low ν (long λ) radio waves through the visible range and into the high ν (short λ) cosmic ray region.

Visible light is nothing more than that part of the electromagnetic spectrum to which the human eye is sensitive. The visible region is only a small portion of the electromagnetic spectrum, ranging from about 4×10^{-7} m (400 nm) to 7×10^{-7} m (700 nm).

Key 14 Wave-particle duality

OVERVIEW *In the early twentieth century Albert Einstein and Max Planck proposed that electromagnetic waves could be quantized. In 1924, deBroglie suggested that waves could be considered as particles and that particles could be treated as waves.*

In this manner, light can be considered as a stream of massless particles called **photons**, each containing a packet or **quantum** of energy. The energy associated with each quantum is given by:

$$E = h\nu$$

where h is Planck's constant and ν is the frequency of the radiation.

Planck's constant is the proportionality constant between the frequency and energy of a photon of light. h has the value 6.63×10^{-27} erg-s (or 6.63×10^{-34} J-s).

DeBroglie proposed that the wavelength of a matter-wave should be

$$\lambda = \frac{h}{mv}$$

where λ is the particle or wave. [It is important to keep the variables v (for velocity) and ν (for frequency) clearly defined.]

The term **wave-particle duality** describes the situation where light can be described equally well and simultaneously as if it has wave-like and particulate character. The ambiguity occurs because both formalisms are equally able to describe the behavior of light. We will see that the same phenomenon occurs when describing electrons in atoms.

Theme 5 STOICHIOMETRY

*S*toichiometry is the study of the mass relationships between chemically equivalent quantities. These relationships are expressed as the subscripts in chemical formulas or as the coefficients in balanced chemical equations.

Key 15   Dalton's atomic theory

OVERVIEW *In the first decade of the nineteenth century, John Dalton proposed a series of postulates know as Dalton's atomic theory. This theory was able to rationalize experimentally determined relationships among the masses of the elements in chemical compounds.*

The postulates of **Dalton's atomic theory** can be summarized as follows:

1. The basic unit of an element is an extremely small, indivisible particle called an atom.
2. Atoms are hard spheres with a characteristic size and mass.
3. Each atom of the same element has the same mass.
4. Atoms of one element are different from atoms of another element.
5. Atoms of one element can combine with atoms of a second element, usually in small whole number ratios, to form "compound atoms" or molecules—the characteristic units of a compound.

We now know that the single characteristic that determines the nature of an atom is its atomic number, not its mass. However, at the time it was proposed, Dalton's atomic theory was important because it was able to rationalize the experimentally observed **law of conservation of matter** and **law of constant proportions**. Dalton's theory suggested the **law of multiple proportions** as a way to test the theory.

- The **law of conservation of matter** is the most important law in chemistry and requires that matter is neither created nor destroyed in a chemical reaction. (This law is not quite true since extremely small amounts of matter can be converted to large amounts of energy as shown by Einstein's famous equation

$$E = mc^2$$

- The **law of constant proportions** states that a pure chemical compound has the same percentage composition of the elements no matter where or how the compound is found or made. This law shares equal importance with the law of conservation of matter.
- The **law of multiple proportions** states that certain pairs of elements are known to result in two different compounds. If samples of these two different compounds are selected so that each has the same mass of one of the elements, the ratio of the masses of the other element will be a ratio of small whole numbers.

Key 16 Atomic and formula masses

OVERVIEW *An atom of each element has a characteristic mass and, in like manner, each molecule of a compound has a characteristic formula mass. Chemists measure amounts of materials in moles, the SI unit for amount of substance. Expressing amount of substance in moles gives the chemically equivalent amount of material.*

The **atomic mass** is the weighted average of the masses of all natural isotopes of an element based on carbon-12 having a mass defined as 12.000 atomic mass units (u). One atomic mass unit is equal to 1.66053×10^{-24} g.

The **gram atomic mass** of an element is that mass of an element obtained when the atomic mass is given in units of grams.

The **chemical formula** of a compound is a representation of the elemental composition of that compound. The **subscripts** following each element indicate the **atomic ratio** of each element in the compound.

The **formula mass** (or **molecular mass**) of a compound is the sum of the masses of the atoms in a chemical formula of that compound.

The **gram-formula mass** of a compound is the amount of that compound that has the same mass in grams as the formula mass in atomic mass units.

A **mole** of a substance is defined as the amount of an element or a compound in grams numerically equal to the substance's atomic or formula mass. One mole of a material contains Avogadro's number of particles of that material.

Avogadro's number is the number of atoms, molecules, or formula units per mole (or gram-atom or gram-formula mass, respectively) of a substance. It is numerically equal to 6.022×10^{23}.

One mole of calcium chloride ($CaCl_2$) contains 1 mol of calcium (Ca) atoms and 2 mol of chlorine (Cl) atoms. The formula mass of $CaCl_2$ is calculated as follows:

$$1 \text{ mol Ca} \times 40.08 \text{ g/mol Ca} = 40.08 \text{ g Ca}$$
$$2 \text{ mol Cl} \times 35.45 \text{ g/mol Cl} = 70.90 \text{ g Cl}$$
$$\text{Formula mass } CaCl_2 = 110.98/\text{mol g } CaCl_2$$

This means that 1 mol $CaCl_2$ contains 110.98 g/mol. In addition, 1 mol $CaCl_2$ contains Avogadro's number (6.022×10^{23}) formula units of $CaCl_2$, 6.022×10^{23} atoms of Ca, and 12.044×10^{23} atoms of Cl.

Key 17 Empirical formula

OVERVIEW *In the laboratory, chemists express the composition of matter in terms of an empirical formula based on experimentally gathered data.*

The **empirical formula** is the simplest chemical formula that expresses the relative number of each element in a compound using the smallest possible whole numbers. To calculate an empirical formula:

1. Determine the mass of each element in the compound. If given the percentage composition, assume 100 g of sample and convert percentages directly to grams.
2. Convert grams of each element to moles of each element.
3. Determine the smallest number of moles calculated in step 2. Divide the moles of each element by that number to obtain a normalized list of integers.
4. All numbers in step 3 that are within +/− 0.1 of a whole number should be rounded to the closest integer. If each element now has a whole number of moles, use these integers as subscripts in the formula.
5. (*Use only if needed.*) If step 4 did not result in all integers, determine rational fractions (e.g., $^3/_2$, $^5/_3$) for numbers not rounded to integers. Multiply all mole values by the denominators of the rational fractions to create whole numbers. Use these integers as subscripts in the empirical formula.

Suppose we need to determine the empirical formula of a compound that contains only nitrogen and oxygen and has 25.93% nitrogen and 74.07% oxygen.

We assume that we have a 100-g sample of the compound, in which case we have 25.93% of 100 g or 25.93 g of N. Similarly, we have 74.07 g of O. The moles of each are calculated as

$$\frac{74.07 \text{ g O}}{16.0 \text{ g O}/\text{mol O}} = 4.63 \text{ mol oxygen}$$

$$\frac{25.93 \text{ g N}}{14.0 \text{ g N}/\text{mol N}} = 1.852 \text{ mol nitrogen}$$

Dividing both by 1.852 gives

$$\frac{4.63 \text{ mol O}}{1.852} = 2.5 \text{ mol oxygen}$$

$$\frac{1.852 \text{ mol N}}{1.852} = 1.0 \text{ mol nitrogen}$$

The 2.5 mol of oxygen is the rational fraction $^5/_2$. We multiply both values by the denominator 2 to get 5 mol oxygen and 2 mol nitrogen. The empirical formula is written as N_2O_5.

Key 18   Percent composition

OVERVIEW *The percent composition of a compound expresses the elemental composition of the compound (found by chemical analysis) in terms of the number of grams of each element divided by the total number of grams present.*

The **percent composition** of a compound is an expression of the elemental composition of that compound. The percent composition of each element (actually the mass percent) is equal to the mass of that element divided by the total mass present (the formula mass) multiplied by 100%:

$$\% \text{ composition A} = \frac{\text{grams A}}{\text{total grams}} \times 100\%$$

Note that the sum of the percent (by mass) of each element in a compound equals 100%.

To determine the percent composition from the formula of a compound, first determine the amount (in g) of each element in one mole of that compound by multiplying the relative number of moles by the atomic mass of the element. Then, by dividing this by the total number of grams in one mole of the compound (the formula mass), you can find the mass percent of each element. For example, to calculate the percent composition of $Co_3(PO_4)_2$, note that each formula unit contains

$$3 \text{ mol Co} \times 58.93 \text{ g/mol Co} = 176.79 \text{ g Co}$$
$$2 \text{ mol P} \times 30.97 \text{ g/mol P} = 61.94 \text{ g P}$$
$$8 \text{ mol O} \times 16.00 \text{ g/mol O} = 128.00 \text{ g O}$$

These add up to 366.73 g/mol for the formula mass of $Co_3(PO_4)_2$.

To calculate the percent composition, divide the mass of each element in one mole by the formula mass (in g/mol):

$$\text{mass } \% \text{ Co} = \frac{176.79 \text{ g Co}}{366.73 \text{ g}} \times 100\% = 48.18\% \text{ Co}$$

$$\text{mass } \% \text{ P} = \frac{61.94 \text{ g P}}{366.73 \text{ g}} \times 100\% = 16.89\% \text{ P}$$

$$\text{mass } \% \text{ O} = \frac{128.00 \text{ g O}}{366.73 \text{ g}} \times 100\% = 34.90\% \text{ O}$$

Theme 6 CHEMICAL EQUATIONS

A chemical equation is a chemical shorthand notation that describes a reaction between chemical substances and provides the necessary information to calculate weight relationships in the reaction. Typically, a chemical reaction is accompanied by the absorption or emission of energy in the form of heat.

Key 19   The equation

OVERVIEW *A chemical equation allows a chemist to quickly determine the quantitative relationships among reactants and products in a chemical reaction. When there is conservation of both mass and charge, an equation is said to be balanced. In that case, the arrow indicates the direction of the reaction and is equivalent to an equal sign.*

In general, an equation is of the following form: **reactants → products**.

Reactants are the species undergoing reaction.

Products are the species being produced in the reaction.

A chemical equation must satisfy the following three conditions:

1. It must be consistent with the "chemical facts."
2. It must retain conservation of mass.
3. It must retain conservation of charge.

The first condition means that all the reactants and products that are observed must be included in the equation.

The following rules help in **balancing** simple chemical equations **by inspection**. More-involved techniques apply when balancing more complex equations, such as those describing oxidation-reduction (or redox) reactions.

1. Balance the elements in the most complicated formula (the one with the most atoms) first, one at a time, using fractions if necessary.
2. Always balance uncombined elements (elements occurring alone or in molecules such as O_2 or N_2) last.
3. Delay balancing elements that occur in more than one compound for next to last.
4. If necessary, clear fractions (by multiplying through by the proper number) or simplify fractions (by dividing through by the lowest common denominator).
5. Never add any other reactant or product to balance the equation or change a subscript once the equation is written. If all the elements do not balance at the end, start over.

The **physical state** of the reactants and products in a chemical equation are indicated by placing (*s*), (*l*), (*g*), or (*aq*) after the chemical formula to indicate solid, liquid, gas, or aqueous solution, respectively.

Key 20 Stoichiometry calculations

OVERVIEW *A balanced chemical equation represents the stoichiometry of the reaction—the relative number of moles of reactants that react to form products.*

The **coefficients** in a balanced equation give the relative number of moles of reactants and products in that equation.

Molar ratios between any pair of chemical species in a particular reaction can be written in order to convert an amount of one quantity to another.

As an example of a calculation using chemical equations, consider the observation that methane gas (CH_4) burns (reacts) in oxygen gas (O_2) to form carbon dioxide (CO_2) and water (H_2O). The unbalanced equation for this reaction is

$$CH_4(g) + O_2(g) \rightarrow CO_2(g) + H_2O(l)$$

To balance the equation, note that each CH_4 molecule contains four H atoms, so that a 2 must be placed in front of the water molecule on the right to balance the four H atoms. There are then a total of four O atoms on the right (two from the CO_2 and one from each of the two water molecules). Placing a 2 before the O_2 on the left balances the equation:

$$CH_4(g) + 2O_2(g) \rightarrow CO_2(g) + 2H_2O(l)$$

The balanced chemical equation provides mole ratios that can be used as factor labels in calculations. Some mole ratios from the balanced equation above are

$$\left(\frac{1 \text{ mol } CH_4}{2 \text{ mol } O_2}\right) \quad \left(\frac{1 \text{ mol } CH_4}{2 \text{ mol } H_2O}\right) \quad \left(\frac{1 \text{ mol } CH_4}{1 \text{ mol } CO_2}\right)$$

$$\left(\frac{2 \text{ mol } O_2}{1 \text{ mol } CO_2}\right) \quad \left(\frac{2 \text{ mol } O_2}{2 \text{ mol } H_2O}\right) \quad \left(\frac{1 \text{ mol } CO_2}{2 \text{ mol } H_2O}\right)$$

Six additional mole ratios are the inverse of those shown.

To calculate the number of moles of O_2 gas that will react with 4.30 moles of CH_4 gas according to the above reaction, realize that two moles of O_2 react with one mole of CH_4, so that there must be twice the number of moles of O_2 as the number of moles of CH_4. The mathematical calculation is as follows:

$$4.30 \text{ mol } CH_4 \times \frac{2 \text{ mol } O_2}{1 \text{ mol } CH_4} = 8.60 \text{ mol } O_2$$

Key 21 Limiting reagent, theoretical and percent yield

OVERVIEW *A balanced chemical equation allows one to calculate the masses of chemically equivalent quantities. One can then determine the quantities of reactants used and products formed in that particular reaction.*

The **coefficients** in a balanced chemical equation represent the relative numbers of moles of reactants and products entering into that reaction.

These coefficients can be used to set up **mole ratios** that relate chemically equivalent amounts of any pair of reactants and/or products in the reaction described by the equation. A particular mole ratio holds for only that reaction.

The **mole ratios** are then used as conversion factors in setting up calculations involving the equation.

If the reactants are present in exactly the right ratio (or **stoichiometry**), then each reactant will be completely consumed when the reaction is complete. The **limiting reagent** is defined as the reactant or reagent that disappears first when two or more reactants undergo reaction. The limiting reagent therefore determines the amount of product formed.

The **theoretical yield** is the maximum amount of product that can be produced in a given reaction under ideal conditions.

The **percent yield** is the yield obtained in a reaction and is expressed as a percentage of the theoretical yield. Percent yield can be calculated by using the following equation:

$$\% \text{ yield} = \frac{\text{actual yield}}{\text{theoretical yield}} \times 100\%$$

Key 22 Enthalpy changes

OVERVIEW *Both chemical and physical processes can occur with the absorption or evolution of energy, usually in the form of heat. In this Key, we will consider only these changes. Other thermodynamic terms are discussed in Theme 15.*

A **calorimeter** is a device used to measure energy changes associated with a physical or a chemical process.

The **heat capacity** of a substance is the amount of heat necessary to raise the temperature of that substance by 1°C.

The **specific heat** of a substance is the amount of heat needed to raise the temperature of one gram of that substance by 1°C.

Enthalpy (*H*) is a measure of the heat content of a substance.

The **heat of reaction (Δ*H*)** is the enthalpy change associated with that reaction. Heat of reaction can be calculated by using the following equation:

$$\Delta H_{reaction} = (\Sigma H \text{ products}) - (\Sigma H \text{ reactants})$$

If a reaction takes place with the evolution of heat, Δ*H* is negative and the reaction is said to be **exothermic**. If a reaction takes place with the absorption of heat, Δ*H* is positive and the reaction is said to be **endothermic**.

The **heat of combustion** of a compound is the amount of heat evolved in the reaction of that compound with oxygen.

The physical processes of melting and boiling also have the following associated enthalpy changes:

• The **heat of fusion (Δ*H_f*)** is the amount of heat necessary to convert one mole of a solid to one mole of liquid.
• The **heat of vaporization (Δ*H_v*)** is the amount of heat necessary to convert one mole of a liquid to one mole of gas.

The **standard heat of formation (Δ*H_f°*)** of a compound is the enthalpy change when one mole of that compound in its **standard state** (1 atmosphere of pressure and 25°C) is formed from its elements in their standard states.

The Δ*H_f°* for any **element in its standard state** is defined as zero.

The **standard heat of reaction** ($\Delta H°$) is the heat released or absorbed when the moles of reactant specified in the balanced chemical equation react to produce the specified moles of product. The units for $\Delta H°$ are usually kJ. The standard heats of formation can be combined using the equation below to obtain the standard heat of reaction.

$$\Delta H°_{\text{reaction}} = \sum (\Delta H°_f \times \text{no. of moles})_{\text{products}} - \sum (\Delta H°_f \times \text{no. of moles})_{\text{reactants}}$$

Note that each heat of formation must be multiplied by the number of moles of each reactant and product specified in the balanced chemical equation.

Hess's law states that when thermochemical equations are added, the heats of reaction are added to give the net heat of reaction.

KEY PROBLEM 1

The use of thermochemical data is illustrated by the following examples:

Methyl alcohol, originally made by the destructive distillation of wood, is now produced according to the reaction

$$CO_2(g) + 2H_2(g) \rightarrow CH_3OH(l)$$

From the standard heats of formation of $CH_3OH(l)$ (-238.7 kJ/mol) and $CO(g)$ (-110.5 mJ/mol), calculate the standard enthalpy change ($\Delta H°$) for this reaction.

From the equation relating the standard heat of reaction to standard heats of formation and the fact that $\Delta H°_f$ for an element in its standard state is zero (H_2 in this example), then

$$\Delta H°_{\text{reaction}} = \sum (\Delta H°_f \times \text{no. of moles})_{\text{products}} - \sum (\Delta H°_f \times \text{no. of moles})_{\text{reactants}}$$

$$= [(-238.7\,\text{kJ/mol}\,CH_3OH)(1\,\text{mol}\,CH_3OH)] - [(-110.5\,\text{kJ/mol}\,CO_2)(1\,\text{mol}\,CO_2)]$$

$$= -128.2\,\text{kJ}$$

KEY PROBLEM 2

Hess's law can be used to calculate the standard heat of formation of a compound, given a series of reactions that can be algebraically manipulated to give the equation for the formation of that compound from its component elements. For example, ΔH_f° for $C_3H_8(g)$ can be determined from the following data:

$$C_3H_8(g) + 5\,O_2(g) \rightarrow 3\,CO_2(g) + 4\,H_2O(g) \quad \Delta H^\circ = -2225\,\text{kJ}$$
$$C(s) + O_2(g) \rightarrow CO_2(g) \quad\quad\quad\quad\quad\quad \Delta H^\circ = -393.7\,\text{kJ}$$
$$2\,H_2(g) + O_2(g) \rightarrow H_2(g) \quad\quad\quad\quad\quad \Delta H^\circ = -569.8\,\text{kJ}$$

Rearrange the first equation to make C_3H_8 the product, remembering that reversing the equation changes the sign of ΔH. Next, by using the appropriate multipliers, use the remaining equations to cancel out all other compounds, leaving only $C(s)$ and $H_2(g)$ on the reactant side. When multiplying an equation by a factor, you must also multiply the relevant ΔH values.

$$3\,CO(g) + 4\,H_2O(g) \rightarrow C_3H_8(g) + 5\,O_2(g) \quad \Delta H^\circ = +2225\,\text{kJ}$$
$$3\,C(s) + 3\,O_2(g) \rightarrow 3\,CO_2(g) \quad\quad\quad\quad\quad \Delta H^\circ = 3(-393.7)\,\text{kJ}$$
$$4\,H_2(g) + 2\,O_2(g) \rightarrow 2\,H_2(g) \quad\quad\quad\quad\quad \Delta H^\circ = 2(-569.8)\,\text{kJ}$$

Adding these equations gives the expression for the formation of $C_3H_8(g)$ from the elements:

$$3\,C(s) + 4\,H_2(g) \rightarrow C_3H_8(g) \quad\quad\quad\quad \Delta H_f^\circ = +95.7\,\text{kJ/mol}$$

Theme 7 ATOMIC STRUCTURE

*D*uring the late nineteenth and early twentieth centuries, several discoveries were made about the nature of the atom. The results of these discoveries led to a model for the atom that was quite different from one that could be visualized on the basis of classical concepts.

Key 23 Historical background

OVERVIEW *By the early decades of the twentieth century, several critical experiments had been performed that led to our modern understanding of the atom.*

Toward the end of the nineteenth century, physical chemists were investigating a phenomenon called **line emission spectra**, the emission of radiation at specific wavelengths and frequencies by atoms that had been energized (excited).

The **atomic emission spectrum** of an atom was found to be a set of frequencies characteristic of that atom and independent of its chemical state.

Furthermore, an atom was found to absorb energy at exactly the same frequencies (**atomic absorption spectrum**) as those found in its emission spectrum. (See Key 13 for the relationship between frequency and energy.)

A simple relationship was derived by **J. Rydberg**, who was able to account for the frequencies of all the lines in the spectrum of hydrogen using a single constant (the **Rydberg constant**).

In the 1890s, **J. J. Thomson** investigated the properties of **cathode rays**. He found that cathode rays consist of a stream of electrons emitted by the cathode (negative electrode) in a gas discharge (or cathode ray) tube. By experimenting with this device, Thomson was able to measure the **charge-to-mass ratio** of the electron.

On the basis of his experiments, Thomson proposed what has become known as the **plum pudding model** of the atom. Thomson proposed that the atom was a positively charged sphere of matter in which electrons were embedded like raisins in a plum pudding.

Robert Millikan performed an **"oil drop" experiment** and determined the charge on the electron. This work also allowed calculation of the mass of the electron from the known charge-to-mass ratio.

The results of **Rutherford's** experiment—bombarding thin sheets of gold foil with alpha particles (helium nuclei, He^{2+})—led to a new model of the atom. This new model showed a high concentration of mass and positive charge at the nucleus of the atom with the very light electrons distributed throughout most of its volume.

Key 24 The modern atom

OVERVIEW *From a chemist's point of view, an atom contains three species: neutrons, protons, and electrons.*

The **nucleus** of an atom is the very small, dense, positively charged central portion of the atom containing protons and neutrons (and therefore most of the mass of the atom).

Neutrons are neutral particles that are part of the nucleus of the atom and have a mass of 1.6749×10^{-24} g.

Protons are positively charged particles that are part of the nucleus of the atom and have a mass of 1.6726×10^{-24} g and a charge of 1.602×10^{-19} coulomb.

Electrons are negatively charged particles that occupy most of the volume of the atom and have a mass of $9.1093897 \times 10^{-28}$ and a charge of -1.602×10^{-19} coulomb.

The **diameter of the nucleus** is approximately 10^{-14} m (0.01 pm), while the **diameter of an atom** is approximately 10^{-10} m (100 pm). In terms of volume, the nucleus occupies approximately $1/10^{12}$ of the atom's volume. The remaining volume is occupied by the electrons.

The **diameter of the nucleus** is on the order of 10^{-12} cm. The **diameter of the entire atom** is on the order of 10^{-8} cm.

The **atomic mass number** (A) of an atom is equal to the number of protons plus the number of neutrons in the nucleus.

The **atomic number** (Z) of an atom is equal to the number of protons in the nucleus. Z uniquely determines the identity of a particular element.

Each atom is characterized by a value of A and of Z. Normal chemical notation uses a **symbol** to designate a chemical element. The atomic number Z is at the lower left of the symbol, and the atomic mass number A is at the upper left:

$$_{Z}^{A}Q$$

Isotopes are atoms of a given element that have the same number of protons in the nucleus (same Z) but different numbers of neutrons (different A); therefore, they differ in atomic mass.

The **atomic mass** is the mass of an element in atomic mass units (u), where one atomic mass unit is equal to $^{1}/_{12}$ the mass of an atom of ^{12}C. ($1\,u = 1.66053 \times 10^{-24}$ g.)

Key 25   The Bohr model of the atom

OVERVIEW *Neils Bohr proposed the first modern model of the atom that overcame the restrictions of classical mechanics and paved the way for the development of modern quantum mechanics.*

In light of the experiments described in Key 23, Bohr proposed the **solar system model** of the atom. In this model the negatively charged electrons travel in circular orbits around the central positively charged nucleus in the same manner that the planets orbit the sun.

Bohr proposed that electrons traveled in distinct orbits and that there were only certain **allowed orbits**. These allowed orbits were constrained by the quantum condition that their **allowed energies** must obey the relationship $E_n = -A/n^2$, where A is a constant and n is an integer denoting the orbit. $n = 1$ represents the orbit closest to the nucleus (in the hydrogen atom this is the **ground state** of the electron), and increasing numbers indicate orbits further from the nucleus.

The radius of the orbits in the Bohr atom is calculated as $n^2(53\text{ pm})$. The radius of the first orbit is 53 pm, which is often called the **Bohr radius**.

Bohr's theory stated that electrons could move from one orbit to another. Energy had to be added to excite an electron to an orbit with a larger value of n (an **excited state**). Energy was released when an electron dropped to an orbit with a lower value of n. The change in energy was

$$E_{\text{final}} - E_{\text{initial}} = \frac{-A}{n_{\text{final}}^2} - \frac{-A}{n_{\text{initial}}^2}$$

$$\Delta E = -A\left(\frac{1}{n_{\text{final}}^2} - \frac{1}{n_{\text{initial}}^2}\right)$$

The negative sign indicates a release of energy.

The constant A duplicated to an unprecedented precision the Rydberg constant that was used to empirically describe line spectra in an equation with the same format. This and other aspects made Bohr's theory one of the most widely and quickly accepted scientific theories.

Key 26 The wave mechanical picture
of the atom

OVERVIEW *The behavior of electrons in atoms is best viewed in the formalism of quantum mechanics, in which electrons are considered to be wavelike in character and able to occupy only certain energies in an atom.*

In 1924, **deBroglie** proposed that electrons could be considered as both waves and particles (Key 14), thereby allowing use of the formalism of wave mechanics to describe the behavior of electrons in atomic systems.

Therefore, one speaks of the **wave mechanical** picture of atomic structure that considers the electron to have wavelike character.

The **Schrödinger equation** is a partial differential equation describing the total energy of electronic systems.

A **wave function** (ψ) is a mathematical expression that is a solution to the Schrödinger equation and describes the behavior of an electron in an atom.

The **square of the wave function** (ψ^2) is a **probability function**; that is, it gives the probability of locating an electron in a given volume of space.

Heisenberg's uncertainty principle is a fundamental limitation on the measurement of certain pairs of observables, such as position and momentum. As a result, it is impossible to determine both the position and energy of an electron in an atom, making it more acceptable to describe the behavior of an electron in an atom in terms of probabilities.

An **electronic orbital** is the region in space around a nucleus in which there is maximum probability of finding an electron.

Key 27 Quantum numbers

OVERVIEW *Solutions to the Schrödinger equation contain four constants called quantum numbers. These quantum numbers describe the behavior of electrons in atoms.*

The four **quantum numbers** are constants that designate the energies and locations of electrons in electronic orbitals.

The **principal quantum number** (n) specifies an electron's energy level and determines the size of the orbital; it can have only the values 0, 1, 2, 3, . . . , n. (The principal quantum number roughly corresponds to the integer n in Bohr's equations.)

The **azimuthal (or angular momentum) quantum number** (l) can have values of 0, 1, 2, 3, . . . , ($n - 1$) and indicates the type of electronic orbital. Values of these quantum numbers correspond to s, p, d, and f orbitals for $l = 0$, 1, 2, and 3, respectively.

The **magnetic quantum number** (m_l), which can have values from $-l$ to $+l$, denotes the spatial orientation of an orbital.

The **spin quantum number** (m_s) can have a value of $+1/2$ or $-1/2$ and is the quantum number that determines the direction in which an electron seems to be spinning. The property of **electron spin** results from the fact that the electron can appear to act as a tiny spinning sphere of negative charge that thereby generates a magnetic field.

The **Pauli exclusion principle** states that no two electrons in the same atom can have the same four quantum numbers.

Note the following:

- One quantum number (n) defines an **energy level**.
- Two quantum numbers (n, l) define a **sublevel**.
- Three quantum numbers (n, l, m_l) define an **orbital**.
- Four quantum numbers (n, l, m_l, m_s) define an **electron**.

Key 28   The electronic structure of atoms

OVERVIEW *By using a set of simple rules, the electronic configuration of each element can be described.*

The **aufbau principle** is used to build up the electronic configurations of the elements. As protons are individually added to the nucleus to build up the elements, electrons are similarly added to the atomic orbitals.

The **electronic configuration** of an atom is the distribution of electrons in the various energy levels in that atom.

Hund's rule states that, given the choice, there is a maximum number of unpaired electrons in an electronic configuration. This means that when faced with two or more **degenerate orbitals** (electronic orbitals having the same energy), electrons go into separate orbitals with parallel spins (same value of m_s).

An **energy level diagram** (or electronic configuration) is a figure showing the electronic orbitals found at different energies. It presents the electronic orbitals in order of increasing energy:

$1s < 2s < 2p < 3s < 3p < 4s < 3d < 4p < 5s < 4d < 5p < 6s < 4f < 5d < 6p < 7s < 5f$

Manganese (Mn), with $Z = 25$, has an electronic configuration of $1s^2 2s^2 2p^6 3s^2 3p^6 4s^2 3d^5$.

The periodic table is a useful reference for constructing an energy level diagram. Reading the periodic table from left to right indicates the orbitals in the correct filling order.

Theme 8 THE PERIODIC TABLE

*T*he periodic table displays the known chemical elements in such a manner that relationships among their physical and chemical properties become evident. Several simple properties of the elements are used to illustrate these trends.

Key 29 Historical basis

OVERVIEW *As more and more elements were discovered, attempts were made to classify them in terms of their chemical and physical properties. This eventually led to the Mendeleev periodic table, which diagrammatically illustrates these relationships for all known elements.*

As early as 1817, **Dobereiner** described a numerical relationship among the atomic weights of certain **triads**—groups of three elements having similar properties. For example, lithium, sodium, and potassium are chemically similar. The average of the atomic masses of Li (7) and K (39) equals the atomic mass of Na (23).

In 1866, **Newland** proposed a **law of octaves**, the first partially successful systematization of the elements that indicated elements with similar properties tended to appear at regular intervals of eight when listed in order of increasing atomic mass. This scheme was not entirely consistent and lacked predictive value.

In 1869, **Mendeleev** proposed a **periodic law** that encompassed all the known elements. By observing systematic variations in physical properties such as melting and boiling points, he was able to state that the properties of the elements are periodic functions of their atomic weights. A later revision arranged the elements in order of increasing atomic number and became known as **the periodic law**.

Key 30 The modern periodic table

OVERVIEW *The modern periodic table is an arrange-*
ment of the elements in order of increasing atomic number.
This arrangement illustrates similar electronic configura-
tions and trends in chemical and physical properties.

The **periodic table** consists of 7 periods (rows) and 32 groups
(columns). For clarity, the 14 columns of the lanthanide and actinide
elements are usually removed and placed at the bottom of the table.
The remaining 18 columns are numbered 1 through 18. In the
older numbering system columns 1 and 2 are numbered IA and IIA.
Columns 13 through 17 are numbered IIIA to VIIA, and the noble
gases are labeled as group O. Columns 3 through 12 are labeled as B-
group elements.

The periodic table is divided in two by a zigzag line called the **Zintl
border** that separates the metallic elements from the nonmetallic
elements.

Metals are elements that are lustrous, have high thermal and electrical
conductivity, and are generally ductile and malleable (easily worked).
They are located to the left of the Zintl border in the periodic table.

Nonmetals are elements that are located to the right of the Zintl border
and tend to be brittle, dull-appearing, and nonconductive. Often they
are gases at room temperature.

Metalloids are elements that have properties intermediate between those
of nonmetals and metals and are located directly on either side of the
Zintl border.

Representative elements are elements in Groups IA through VIIA (or,
in a new designation, Groups 1, 2, and 13 through 17).

Transition metals (or **transitional elements**) are elements in Groups IB
through VIIIB (or Groups 3 through 12).

A **period** is a horizontal row of elements in the periodic table.

A **group** (or **family**) is a vertical column of elements in the periodic
table:

- The **alkali metals** are elements in Group IA (or 1).
- The **alkaline earth metals** are elements in Group IIA (or 2).
- The **pnictogens** (or **pnictides**) are elements in Group VA (or 15).

- The **chalcogens** (or **chalcogenides**) are elements in Group VIA (or 16).
- The **halogens** (or **halides**) are elements in Group VIIA (or 17).
- The **noble gases** are elements in Group VIIIA (or 18).

The **actinide** (or **transuranium**) elements are the elements located after uranium (actually thorium through lawrencium).

The **rare earths** (or **lanthanides**) are the elements cerium through lutetium.

The shape of the periodic table is a result of filling the various energy levels:

- Groups IA and IIA (1 and 2) fill the s sublevel.
- Groups IIIA through VIII (13 through 18) fill the p sublevel.
- The transition elements fill the d sublevel.
- The lanthanides and actinides fill the f sublevel.

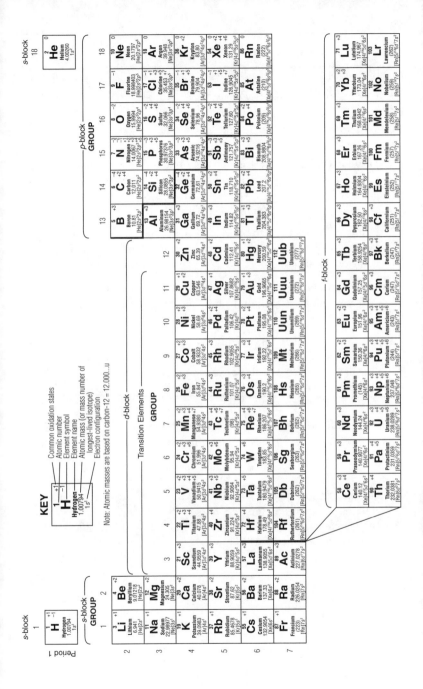

Key 31 Periodic trends

OVERVIEW *Several physical properties of the elements exhibit definite periodic trends, that is, size, ionization energy, and electron affinity.*

The **atomic radius** is the effective radius of an atom measured in its elemental state (one-half the interatomic distance). Atomic radii tend to increase going down a column in the periodic table and to decrease going across a row (from left to right).

The **ionic radius** is the size of an ion—a charged species (Key 32). For ions of the same charge, that is, in the same family, radius increases going down the column.

Isoelectronic species are species (atoms, ions, or molecules) that have the same number of electrons. For an isoelectronic series of ions, the radius decreases with increasing atomic number.

The **ionization energy** (or **ionization potential**) is the energy needed to remove an electron from an isolated atom. The trends for ionization energies are inverse to those for atomic radii; that is, the ionization energies decrease going down a column in the periodic table and increase going across a row.

The **electron affinity** is the energy emitted in the process of adding an electron to an atom (forming a negative ion). Elements with the highest electron affinities lie toward the upper right corner of the periodic table.

Diagonal relationships are properties that vary regularly from one extreme at the lower left corner to the other extreme at the upper right corner. For example, ionization energies are lowest for the element francium and highest for fluorine.

Theme 9 THE CHEMICAL BOND

*A*toms combine to form compounds. In general, the study of chemical bonds involves consideration of the forces that link atoms together. In a broad sense, bonding can be considered to be either ionic or covalent, representing two extremes in a continuity of bond types. A theory of bonding should help to rationalize the shapes of molecules as well as their stoichiometries.

Key 32 Ionic bonding

OVERVIEW *The ionic bond (one of the two extreme forms of bonding) involves the transfer of an electron from an atom of a metal (forming a positively charged species) to an atom of a nonmetal (forming a negatively charged species). These two species then interact electrostatically to form an ionic bond.*

An **ion** is an atom or group of atoms with a positive or negative charge.

A **cation** is a positively charged ion.

An **anion** is a negatively charged ion.

An **ionic bond** is a bond formed between a positively charged ion and a negatively charged ion.

An **ionic bond** is a bond formed by the electrostatic attraction between ions with positive and negative charges.

The strength of the electrostatic interaction is given by **Coulomb's law**:

$$E = k\frac{q_+ q_-}{r}$$

E is the energy of the interaction, q_+ and q_- are the charges on the positive and negative ions, respectively, r is the distance between the ions, and k a constant.

The stoichiometry of an ionic compound is determined by **electroneutrality**; that is, the total number of positive charges must equal the total number of negative charges.

Key 33 Covalent bonding

OVERVIEW *The covalent bond (the other extreme form of bonding) involves the sharing of one or more pairs of electrons by the atoms involved in forming the bond.*

A **covalent bond** is the result of two atoms sharing one, two, or three pairs of electrons in order to achieve an octet of electrons.

Most covalent compounds tend to obey the **octet rule** which states that many atoms, when entering into bonding, tend to attain a **noble gas configuration** (ns^2np^6)—the outer electronic configuration found in the noble gases (a filled sp sublevel with the same value of n for a total of eight electrons).

Valence electrons are electrons in the outer electronic configuration of an atom (outside the noble gas configuration).

A **single bond** is a covalent bond in which one pair of electrons is shared.

A **double bond** is a covalent bond in which two pairs of electrons are shared.

A **triple bond** is an covalent bond in which three pairs of electrons are shared.

A **diatomic molecule** is a molecule composed of two atoms.

A **polyatomic molecule** is a molecule containing more than two atoms.

A **polyatomic ion** is an ion composed of two or more atoms. Typically, the bonding between the atoms in a polyatomic ion is covalent, but the charged ion then interacts with oppositely charged ions to form an ionic compound.

The **bond energy** is the amount of energy needed to separate two atoms joined by a chemical bond.

The **bond length** is the distance between the nuclei of two atoms joined by a chemical bond.

The **bond vibrational frequency** is the frequency at which the atoms vibrate relative to each other.

Key 34 Lewis structures, resonance

OVERVIEW *The stoichiometry of covalently bonded molecules can be viewed by considering their Lewis structures, the distribution of the bonding electrons around the atoms in such a manner as to attain a noble gas electronic configuration.*

1. Count all the valence electrons (outermost *s* and *p* electrons) from all the atoms in the structure. Include extra electrons for each negative charge on an ion and subtract one electron for each positive charge.
2. Arrange the given atoms in a skeleton structure. The least electronegative atom should be in the center of the structure. Hydrogen atoms are always on the periphery.
3. Add a pair of electrons between all the atoms to form a covalent bond.
4. Place the remaining electron pairs in nonbonding positions to create an octet of electrons around all the other atoms.
5. If the central atom has less than eight electrons (and is not Be or B, see below) then move a nonbonding pair from an outer atom into a bonding position, creating a double bond. Repeat until the central atom has an octet of electrons.

Some molecules do not conform to the octet rule. Molecules containing Be or B as the central atom are **electron-deficient**, with less than eight valence electrons.

Molecules in which the central atom is in groups V through VIII and in the third, fourth, or fifth period have **expanded octets** able to hold up to six pairs of electrons (12 valence electrons). Examples of this are PF_5 and SF_6.

The **formal charge** on an atom is the difference between the number of valence electrons in an isolated atom and the number of electrons assigned to that atom in a Lewis structure.

When **resonance** exists, two or more equivalent Lewis structures can be drawn.

A **resonant structure** is one of the equivalent (Lewis) structures that can be drawn for a single molecule.

A **resonant hybrid** is the actual structure of a molecule for which resonant structures can be written.

Key 35   Bond polarity

OVERVIEW *Most bonds between atoms are partially ionic and partially covalent. This leads to a separation of charge and to the development of a dipole moment.*

Electronegativity (EN) is defined as the tendency of an atom to attract a shared pair of electrons in a chemical bond. The electronegativity scale is normalized to elemental fluorine, which is assigned EN = 4.0. Electronegativity increases going from left to right in the periodic table and from bottom to top in a given group.

The difference in electronegativity (ΔEN) between two atoms in a bond is a measure of the amount of ionic character in that bond. ΔEN = 1.7 corresponds to approximately 50% ionic − 50% covalent.

Ionicity is the amount of ionic character in a bond (compared to covalent character).

In general, an **electric dipole** results from a separation of positive and negative charge and has both magnitude and direction.

A **bond dipole** is a dipole formed between two atoms involved in a bond and is caused by unequal sharing of the electrons in the bond, thus making one end of the bond positive and the other negative.

A **dipole moment** is the product of the partial charge on either end of a dipole multiplied by the distance between the charges.

A **polar bond** is a bond formed between atoms of different electronegativity.

A **polar molecule** is a molecule that has a net electric dipole.

A **nonpolar molecule** is a molecule that does not have a net electric dipole.

Key 36 Valence shell electron pair repulsion theory

OVERVIEW *By using the valence shell electron pair repulsion theory, one is able to predict molecular geometry from Lewis structures. In particular, this theory considers the influence of lone pairs of electrons on geometry.*

The **valence shell electron pair repulsion (VSEPR) theory** is used to predict molecular structure based on the tendency of electron pairs in the valence shell to be as far apart as possible.

The **Lewis structure** of a covalently bonded molecule is essential in establishing the number of bonds and lone pairs of electrons that determine structure in the VSEPR theory.

A **bonding pair of electrons** is a pair of valence electrons in a molecule that is involved in bond formation.

A **lone** (or **nonbonding**) **pair of electrons** is a pair of valence electrons in a molecule that is not involved in bond formation.

A simple treatment of the **electrostatic repulsion** of groups of electrons leads to a prediction of the geometry of simple molecules. If A is the central atom and X represents a group of electrons, then the geometry of the molecule based on electronic repulsion alone is

- AX_2—linear
- AX_3—planar triangular
- AX_4—tetrahedral
- AX_5—triangular bipyramidal
- AX_6—octahedral.

Since pairs of electrons involved in bonding are localized in the region between the two atoms involved, they are less effective in repelling each other than lone or nonbonding pairs that are attached to only one atom. As a result, a lone pair of electrons (E) has an effect on the geometry of a molecule since we see only the atoms, not the lone pair. To illustrate, for four groups of electrons:

- AX_4 results in a tetrahedral molecule.
- AX_3E results in a triangular pyramid molecule.
- AX_2E_2 results in a bent molecule.

Key 37   Valence bond theory
and hybridization

OVERVIEW   *The formation of a covalent bond can also be approached by considering the overlap of atomic orbitals on two different atoms that allows for the mutual exchange of electrons. In this manner, a bond is formed as the electrons spend time around both atoms.*

Valence bond theory is a theory of bonding that views a bond as being formed by sharing a pair of electrons between two overlapping atomic or hybrid orbitals.

A **pair of electrons** is two electrons in the same orbital with opposite values of m_s; that is, they have opposite spins.

Hybridization is the process of combining atomic orbitals in an atom to generate a set of new hybrid orbitals.

Hybrid orbitals are orbitals formed by combining two or more atomic orbitals in the same atom. The general rules for forming hybrid orbitals involve the linear combination of atomic orbitals having the proper symmetry. The number of hybrid orbitals obtained must equal the number of atomic orbitals from which they are formed. For example, four sp^3 **hybrid orbitals** are formed by combining one s and three p atomic orbitals.

Hybrid orbitals have a very specific geometric orientation:

- Two sp orbitals are linear.
- Three sp^2 orbitals form a triangle.
- Four sp^3 orbitals form a tetrahedron.
- Five dsp^3 orbitals form a triangular bipyramid.
- Six d^2sp^3 orbitals form an octahedron.

In each of the above examples, the atom to which the orbitals belong is at the center of the polyhedron.

A **sigma (σ) bond** is a bond formed by head-to-head overlap of two orbitals. It is spherically symmetric about the internuclear axis.

A **pi (π) bond** is a bond formed by the sideways overlap of a pair of p orbitals.

Key 38   Molecular orbitals

OVERVIEW *Molecular orbital theory considers a molecule as a collection of nuclei surrounded by a set of electronic orbitals belonging to the whole molecule. These molecular orbitals are constructed from atomic orbitals.*

A **molecular orbital (MO)** is an orbital that extends over the entire molecule. When a pair of atomic orbitals (one from each atom involved in a bond) are combined, one bonding MO and one antibonding MO are formed.

A **bonding molecular orbital** builds up electron density in the region between the nuclei. When filled, bonding orbitals stabilize a molecule.

An **antibonding molecular orbital** concentrates electron density outside the region between the nuclei. As a result, a filled antibonding orbital destabilizes the molecule.

A **sigma (σ) molecular orbital** is an MO in which electron density is spherically symmetric about the internuclear axis—a line joining the two nuclei involved in the bond.

A **pi (π) molecular orbital** is an MO in which electron density lies above and below the internuclear axis.

Delocalized molecular orbitals are not confined to two adjacent atoms in a bond but are spread out over several atoms.

The **bond order** (net number of pairs of bonding electrons) is equal to the number of electrons in bonding MOs minus the number of electrons in antibonding MOs, divided by 2.

Theme 10 GASES

*T*he gaseous state is the simplest state of matter and therefore the easiest to understand and explain theoretically. This is true because, under normal conditions, the molecules making up a gas can be considered as point masses in rapid motion that do not interact with each other. Using this model, the kinetic molecular theory satisfactorily explains the behavior of the gaseous state.

INDIVIDUAL KEYS IN THIS THEME

Key 39   The gaseous state

OVERVIEW *The gaseous state is a totally disordered state of matter in which the molecules move constantly in rapid, random translational motion. The macroscopic properties of gases reflect this microscopic behavior.*

All gases have the following **characteristic properties**:

- Gases are **easily compressed**; that is, if a sample of a gas is squeezed, its volume is easily reduced.
- A gas **expands to fill the available volume**.
- Gases **exert a uniform pressure** on the walls of the vessel in which they are contained.
- Gases mix in all proportions (they are **miscible**).
- There is relatively **rapid diffusion** in gases. (**Diffusion** is the gradual mixing of the molecules of one gas with those of another as a result of molecular motion.)

An **equation of state** is an equation that describes the physical behavior of a state of matter. In the case of a gas, the simplest equation of state contains the following four variables that completely describe the gas:

- V is the **volume** of the container.
- n is the **amount** of gas (the number of moles).
- P is the **pressure** exerted by the gas.
- T is the **temperature** of the gas.

It is important to note that whenever you specify the temperature for a gas, you must use the **absolute temperature** scale (i.e., temperature is expressed in kelvins) (Key 5).

Key 40   The gas laws

OVERVIEW *A number of historical observations (laws) describe the behavior of gases.*

Avogadro's law states that, at constant temperature and pressure, the volume occupied by a gas is proportional to the number of moles of gas present:

$$V_{(T,P)} = kn$$

where k is a proportionality constant and the subscript (T,P) denotes constant temperature and pressure.

This law is sometimes stated as **Avogadro's principle**: Equal volumes of different gases at the same temperature and pressure contain the same number of molecules.

Boyle's law states that the volume of a given amount of gas held at a constant temperature is inversely proportional to pressure:

$$V_{(n,T)} = k(1/P)$$

The simplest definition of an **ideal gas** is one that obeys Boyle's law.

Charles' law states that the volume of a given amount of gas held at constant pressure is directly proportional to the absolute temperature:

$$V_{(n,P)} = kT$$

The **combined gas law** is a combination of Boyle's and Charles' laws, giving the dependence of the volume occupied by a given amount of gas as both temperature and pressure are varied. For a fixed amount of gas (constant n):

$$V_2 = V_1 \left(\frac{P_1}{P_2} \right) \left(\frac{T_2}{T_1} \right)$$

where subscripts 1 and 2 represent initial and final states, respectively.

Gay-Lussac's law states that the pressure exerted by a given amount of gas held at constant volume is directly proportional to the absolute temperature:

$$P_{(n,V)} = kT$$

Dalton's law of partial pressures states that in a closed container, the total pressure is the sum of the partial pressures of the individual gases:

$$P_T = \Sigma P_i$$

where P_T is the total pressure, P_i are the individual partial pressures, and Σ denotes "sum of."

The **partial pressure** of a gas is the pressure that the gas exerts in a mixture of gases. It is equal to the pressure that the gas would exert if it were the only gas present.

KEY PROBLEM 1

As an example of the use of the combined gas law, consider a sample of a gas at 125°C and 745 torr that occupies 125 mL. Calculate the volume that this gas will occupy at 75°C and 1.25 atm.

First convert all temperatures to kelvins (by adding 273° to the temperatures in °C) and convert both pressures to the same units (in this case, we will convert torr to atm: 745 torr × 1 atm/760 torr = 0.980 atm). Then substitute values for the initial and final values into the combined gas law:

$$V_2 = 1.25 \text{ L} \left(\frac{0.980 \text{ atm}}{1.25 \text{ atm}} \right) \left(\frac{348 \text{ K}}{398 \text{ K}} \right) = 0.892 \text{ L}$$

Alternatively, the pressure and temperature factors must be greater than or less than 1. Since in this example the pressure is increasing, the volume must increase and the pressure ratio will be >1 (0.980 atm/ 1.25 atm); also, since the temperature is decreasing, the temperature ratio will be <1 (348 K/398 K). Then, multiply the initial volume by both the pressure and temperature ratios to obtain the final volume, resulting in the same formula as above.

KEY PROBLEM 2

Using the data in problem 1, (a) Will the temperature change cause an increase or a decrease in volume? (b) Will the pressure change cause an increase or a decrease in volume?

These questions can usually be answered by referring to everyday experience and common sense:

(a) We all have had experience with objects shrinking as they cool. Therefore a decrease in volume is expected.

(b) We all have observed that increasing the pressure on an object can make its volume smaller (stuffing an overfull suitcase is an example). Therefore we conclude that the increasing pressure will make the volume smaller.

Key 41   The ideal gas law

OVERVIEW *Several of the gas laws can be combined to derive the ideal gas law, an equation of state describing the physical behavior of an ideal gas.*

By combining Boyle's law, Charles' law, and Avogadro's principle, the **ideal gas law** can be derived. It states that the volume of an ideal gas is directly proportional to number of moles and to temperature and inversely proportional to pressure:

$$V = RnT\left(\frac{1}{P}\right)$$

or, more commonly: $PV = nRT$

The **gas constant (R)** is the proportionality constant appearing in the ideal gas law. It has the numerical value of $0.0821\,L\,atm/mol\,K$. Note that the **units** of the gas constant are very important. When using this value for R,

- **volume** is expressed in **liters (L)**
- **pressure** in **atmospheres (atm)**
- **amount of gas** in **moles (mol)**
- **temperature** in **kelvins (K)**.

Standard temperature and pressure (STP) is the standard temperature and pressure for a gas: $273\,K$ ($0°C$) and $1.00\,atm$.

The **molar volume of an ideal gas** is the volume occupied by one mole of an ideal gas at STP, numerically equal to $22.4\,L/mol$.

KEY PROBLEM 1

The molar volume of an ideal gas at STP can be used to solve the following problem. The reaction between $2\,NO_2$ and H_2 produces NH_3 and H_2O according to the equation

$$2\,NO_2(g) + 7\,H_2(g) \rightarrow 2\,NH_3(g) + 4\,H_2O(g)$$

How many liters of H_2 at STP are needed to react with $0.20\,mol$ of NO_2 to produce NH_3?

The stoichiometry of the balanced equation states that $7\,mol\,H_2$ is equivalent to $2\,mol\,NO_2$. Use this to convert $mol\,NO_2$ to $mol\,H_2$ (see Key 20):

$$0.20 \text{ mol NO}_2 \times \frac{7 \text{ mol H}_2}{2 \text{ mol NO}_2} = 0.70 \text{ mol H}_2$$

Then, using the molar volume of an ideal gas, convert mol H_2 to liters H_2:

$$0.70 \text{ mol H}_2 \times \frac{22.4 \text{ L H}_2}{1 \text{ mol H}_2} = 16 \text{ L H}_2$$

KEY PROBLEM 2

The use of the ideal gas law is illustrated in the following problem. Determine the number of moles of an unknown gas in a sample that, when contained in a 2.00-L vessel, is found to exert a pressure of 1.16 atm at 335 K.

Rearrange the ideal gas law, $PV = nRT$, solve for n, and substitute the values for R, P, V, and T:

$$n = \frac{PV}{RT}$$

$$= \frac{(1.16 \text{ atm})(2.00 \text{ L})}{(0.0821 \text{ L atm}/\text{mol K})(335 \text{ K})} = 0.0844 \text{ mol}$$

KEY PROBLEM 3

Nitrogen oxide (NO), a component of smog, reacts with oxygen and water vapor in air to form nitric acid:

$$4 \text{ NO}(g) + 3 \text{ O}_2(g) \rightarrow 2 \text{ H}_2\text{O}(g) + 4 \text{ HNO}_3(l)$$

If the partial pressure of NO is 3.3×10^{-5} atm, how many grams of HNO_3 can be formed in the volume of air over a city block, say 1.2×10^7 L, at 25°C?

The ideal gas law allows us to determine n, the number of moles of NO gas. Rearranging and substituting values for R, P, V, and T (remembering to convert °C to K) gives

$$n = \frac{(3.3 \times 10^{-5} \text{ atm})(1.2 \times 10^7 \text{ L})}{(0.0821 \text{ L atm}/\text{mol K})(298 \text{ K})} = 16 \text{ mol NO gas}$$

The balanced equation allows us to convert mol NO to mol HNO_3; and, using the formula weight of HNO_3, we convert to g HNO_3:

$$16 \text{ mol NO} \times \frac{4 \text{ mol } HNO_3}{4 \text{ mol NO}} \times \frac{63 \text{ g } HNO_3}{1 \text{ mol } HNO_3} \times 1.0 \times 10^3 \text{ g } HNO_3$$

KEY PROBLEM 4

Another illustration of an ideal gas law calculation is the determination of the formula weight of a compound. A 1.70-g sample of a gaseous compound contained in a 400.-mL container at 25°C exerts a pressure of 560. torr. Using the ideal gas law, determine the number of moles of compound and the formula mass (FM).

Since the value of $R = 0.0821 \text{ L atm/mol K}$, the pressure must be converted from torr to atm (560. torr $\times$ 1 atm/760 torr = 0.737 atm) and the volume from mL to L (400. mL $\times$ 1 L/1000 mL = 0.400 mL). Inserting the values for R, P, V, and T gives

$$n = \frac{(0.737 \text{ atm})(0.400 \text{ L})}{(0.0821 \text{ L atm/mol K})(298 \text{ K})} = 0.0120 \text{ mol}$$

Since n, the number of moles, is equal to g/FM, the formula mass of this compound is

$$FM = \frac{g}{n} = \frac{1.70 \text{ g}}{0.0120 \text{ mol}} = 141 \text{ g/mol}$$

Key 42   The kinetic molecular theory

OVERVIEW *The kinetic molecular theory explains the macroscopic properties of ideal gases in terms of a model based on microscopic behavior.*

The postulates of the **kinetic molecular theory** are

1. The molecules in a gas are **point masses**; that is, their volume is negligible compared to the volume of the container.
2. There are **no attractive forces** between the molecules of a gas.
3. The molecules of a gas are in **continuous rapid, random, straight-line motion**
4. Collisions between the molecules in a gas are **elastic**. This means that there is no net loss of kinetic energy on collision.
5. The average kinetic energy of the molecules in a gas is directly **proportional to the temperature** of the gas.

In terms of the kinetic molecular theory, an **ideal gas** is one in which there are no attractive forces between the molecules and in which the molecules are point masses.

The pressure a gas exerts is caused by **collisions** of the gas molecules with the walls of the container.

Brownian motion is the continuous random motion of particles suspended in a gas or a liquid. (This is good evidence in support of postulate 3 of the kinetic molecular theory.)

Graham's law of diffusion states that the rate of diffusion of a gas is inversely proportional to the square root of its formula mass (FM). The relationship between the rates of diffusion of two gases (A and B) can be expressed in terms of the following ratio:

$$\frac{\text{rate}_A}{\text{rate}_B} = \left(\frac{\text{FM}_B}{\text{FM}_A} \right)^{1/2}$$

FM_A and FM_B are the formula masses of A and B, respectively.

The distribution of the kinetic energies (and velocities, since kinetic energy = $\frac{1}{2}mv^2$) of the individual molecules in a gas follows **Boltzmann statistics**.

A particular **Boltzmann distribution** of kinetic energies is dependent on temperature; the average kinetic energy of a sample of gas increases with increasing temperature.

Key 43 Deviations from ideal behavior

OVERVIEW *At low temperatures and high pressures, the behavior of most gases deviates from ideal behavior because the gas molecules are not point masses and attractive forces exist between the molecules.*

The **Van der Waals equation** is a modification of the ideal gas law in which corrections are made for the volume of molecules and the presence of intermolecular forces. A typical Van der Waals equation is as follows:

$$\left(P + \frac{an^2}{V^2} \right)(V - nb) = nRT$$

a and b are constants independent of P, V, or T but are dependent on the particular gas.

The term $\left(\frac{an^2}{V^2} \right)$ corrects the pressure because there are attractive forces between the molecules. The **corrected pressure** is

$$\left(P + \frac{an^2}{V^2} \right)$$

The term nb corrects the volume because the molecules making up the gas do have a finite volume. The **corrected volume** is $(V - nb)$.

The value of a is a measure of how strongly the molecules attract one another and therefore increases with increasing intermolecular attraction (Key 47).

The value of b is approximately correlated with molecular size. The larger the species, the larger the value of b tends to be.

Theme 11 SOLIDS, LIQUIDS, AND PHASE TRANSITIONS

*T*he other two states of matter are the solid state and the liquid state. Crystalline solids are characterized by long-range order in three dimensions, while liquids have only short-range order. The physical properties of these states of matter depend on the nature of the forces between the species under consideration. The phase diagram of a substance summarizes the regions of stability of the solid, liquid, and gas state for that substance.

Key 44 The solid state

OVERVIEW *Solids can be either crystalline or amorphous. Most solids are crystalline and are characterized by an ordered array of atoms, ions, or molecules that occupy fixed positions. The macroscopic (bulk) properties of solids support this model.*

A **crystalline solid** is a solid having an ordered internal arrangement of atoms, ions, or molecules. These species vibrate about fixed positions in an ordered geometric array.

An **amorphous solid** is a solid that does not have an ordered internal arrangement, for example, a permanently supercooled liquid such as a glass.

The following are macroscopic (bulk) properties of solids:

1. Most solids have **definite geometric forms**, characteristic faces defining the crystal and characteristic angles between the faces. This is reflected in their external **habit**—the shape and form of a crystal.
2. Most crystalline solids exhibit the phenomenon of **cleavage**— separation along specific planes to give fragments with flat surfaces that are dependent on the geometric array and type of bonding—when subjected to external mechanical stress.
3. Solids are **rigid** because the atoms, ions, or molecules in a solid occupy fixed positions.
4. Solids tend to **maintain their volume** because there are strong attractive forces between the component species.
5. Solids exhibit **very slow diffusion** because of the ordered internal arrangement of atoms and the attractive forces between atoms, ions, and molecules. When appreciable diffusion does occur, it is usually the result of **defects** (or mistakes) in the ordered arrangement.
6. Solids are **practically incompressible**; there is little empty space between the species making up solids.

    Daltonides are stoichiometric solids; that is, the mole ratios of the component elements are small whole numbers. **Berthollides** are nonstoichiometric solids; that is, the mole ratios of the component elements are not small whole numbers. A **substitutional solid solution** is a solid solution in which an atom or molecule of the solute replaces an atom or molecule of the solvent in its lattice. An **interstitial solid solution** is a solid solution in which solute particles fit into empty spaces between the particles of the solvent.

Key 45 The structure of solids

OVERVIEW *The atoms, molecules, or ions making up a crystalline solid are in fixed positions forming an ordered three-dimensional array. X-ray diffraction is a powerful tool that can be used to investigate the structure of solids.*

X-ray diffraction involves the scattering of X rays by the array of species comprising a crystalline solid. This scattering (or diffraction) occurs because these species are arranged in ordered arrays in space forming planes of lattice points. The distance between these planes (d) is in the range of 200 to 400 pm, which is the magnitude of the wavelength (λ) of radiation in the X-ray region of the electromagnetic spectrum.

The **Bragg equation**, $n\lambda = 2d \sin \theta$, is used to analyze X-ray diffraction data obtained from crystals; n is an integer, λ is the wavelength of the radiation, d is the spacing between planes of lattice points, and θ is the angle between the beam of incident radiation and the plane of lattice points.

A **lattice point** is the center of the specific positions occupied by the atoms, molecules, or ions in a crystalline solid.

The **crystal lattice** is a repeating pattern of atoms, molecules, or ions in a crystal, that is, the three-dimensional array lattice points.

The **unit cell** is the smallest part of a lattice that can be repeated over and over in all directions to give the entire crystal lattice. If we define the edges of the unit cell as a, b, and c, and the angles between them as α, β, and γ (α is the angle between edges b and c, and so on), then the seven simple unit cells are as follows:

- **cubic:** $a = b = c$; $\alpha = \beta = \gamma = 90°$
- **tetragonal:** $a = b \neq c$; $\alpha = \beta = \gamma = 90°$
- **orthorhombic:** $a \neq b \neq c$; $\alpha = \beta = \gamma = 90°$
- **rhombohedral:** $a = b = c$; $\alpha = \beta = \gamma \neq 90°$
- **monoclinic:** $a \neq b \neq c$; $\alpha = \gamma = 90°$, $\beta \neq 90°$
- **triclinic:** $a \neq b \neq c$; $\alpha \neq \beta \neq \gamma \neq 90°$
- **hexagonal:** $a = b \neq c$; $\alpha = \beta = 90°$, $\gamma = 120°$

Key 46 The liquid state

OVERVIEW *The liquid state is that state of matter inter-mediate between the gas state and the solid state, in terms of both physical properties and the model used to describe it.*

Liquids are characterized by relatively weak bonds between molecules such that the molecules are free to move around one another. Typically, liquids exhibit **short-range order** as a result of these weak interactions which are not strong enough to give the long-range order characteristic of crystalline solids.

The following characteristic properties of liquids are intermediate between those of gases and solids:

1. Liquids are **relatively incompressible**, indicating that there is little free volume in the liquid state.
2. Although liquids **maintain relatively definite volumes**, they have **no characteristic shape**. Because liquids are not rigid like solids, molecules are free to move over one another. As a result, liquids, under the influence of gravity, conform to the shape of the bottom of a container.
3. Liquids exhibit **relatively slow diffusion**, indicating that the average distance between collisions is small (about 0.3 nm, as compared with about 10 nm in gases).
4. Unlike gases, liquids exhibit a **surface tension** that results from attractive forces trying to pull molecules from the surface into the bulk of the liquid. Surface tension is a measure of the strength of the internal attractive forces in a liquid and increases as these forces become stronger.
5. Liquids have a unique property; they **evaporate from an open container** at temperatures below the boiling point.

Evaporation is the process by which the molecules having the most kinetic energy escape from the surface of a liquid.

The **equilibrium vapor pressure** is the pressure exerted by a vapor in equilibrium with its liquid. It is an example of a **dynamic equilibrium** in the case where the rate of evaporation is equal to the rate of condensation.

The **normal boiling point** of a liquid is the temperature at which the vapor pressure of the liquid is equal to one standard atmosphere (760 torr).

Key 47 Intermolecular forces

OVERVIEW *The basic building block in molecular compounds is the molecule, which is formed by relatively strong covalent bonds. Molecules are then held together in the solid and liquid states by weaker forces that have an effect on the physical properties of the material.*

Intramolecular forces are the forces between the atoms that make up molecules; they are generally considered to be covalent chemical bonds.

Intermolecular forces are forces between the molecules of molecular substances that tend to be weaker than normal covalent bonds. There are several types of intermolecular forces:

1. **Ion-dipole interactions** are intermolecular forces between an ion and a molecule having a dipole—a polar molecule (Key 35). An **induced dipole** is a dipole created when the electron cloud is distorted by a neighboring dipole or ion.
    Polarizability is a measure of the extent to which the electron cloud in an atom or molecule can be distorted (or polarized).
2. **Dipole-dipole interactions** are weak intermolecular forces between polar molecules.
3. **London (dispersion) forces** are weak intermolecular forces between instantaneous dipoles on nonpolar molecules.
    An **instantaneous dipole** is a dipole created by the motion of electrons in an atom.
4. **Hydrogen bonds** are intermolecular forces resulting from a specific interaction between a hydrogen atom covalently bonded to a nitrogen, oxygen, or fluorine in one molecule and an available electron pair on a nitrogen, oxygen, or fluorine on another molecule. These relatively strong intermolecular forces result in the anomalously high melting and boiling points of water and ammonia. These forces are also responsible for important interactions in biologically important molecules.

Key 48 Classification of solids

OVERVIEW *Solids can be classified according to the nature of the bonding between the component species. Each type of solid has certain general physical properties reflecting this bonding.*

Ionic solids are solids in which the building blocks are positive and negative ions. Ionic solids generally have high melting and boiling points and do not conduct heat or electricity. They are formed between a metallic element (from the left-hand side of the periodic table) that forms a cation and a nonmetallic element (from the right-hand side of the periodic table) that forms an anion. Examples are NaCl, MgF_2, and Al_2O_3. Ionic solids are brittle and shatter readily when a force causes positive and negative charges to align within the crystal. This requires a movement of only one ionic diameter.

Molecular solids are solids composed of covalently bonded molecules that are weakly bonded together by intermolecular forces (Key 47). They have relatively low melting and boiling points and, in general, do not conduct electricity. Molecular solids are usually formed from compounds composed of nonmetals. Examples are CCl_4, SF_6, and C_6H_6 (benzene).

Covalent network solids are solids in which atoms form strong covalent bonds in an infinite three-dimensional network of covalently bonded species. Generally, they have high melting and boiling points and may be **semiconductors**—substances that are weak conductors of electricity. Covalent network solids are formed between elements around the Zintl border. Examples are SiC, BN, and AlP.

Metals are solids formed by metallic elements. They tend to have variable melting and boiling points. The **electron-sea model of metallic bonding** is a model in which electrons move throughout an ordered array of metal atoms. The metal ions are visualized as being embedded in a sea of electrons. This arrangement leads to good electrical and thermal conductivity. **Malleability** is the ability of a metal to be hammered or rolled into thin sheets. Unlike the situation in ionic solids, movement of the atoms in the metallic crystal does not result in destructive repulsive forces.

Key 49   Phase diagrams

OVERVIEW *The regions of stability of the solid, liquid, and gas phases for a particular pure material are summarized in a phase diagram for that material. A phase diagram is constructed from careful measurements of melting and boiling points as a function of pressure.*

A **heating curve** is a plot of temperature versus time when adding heat to a closed system at a constant rate. By preparing heating curves for different pressures, one can determine the pressure dependence of the phase transition.

A **cooling curve**, the reverse of a heating curve, is a plot of temperature versus time when removing heat from a closed system at a constant rate.

A **phase diagram** is a plot of pressure versus temperature (P versus T) showing the conditions under which a given phase of a substance exists. Data for a phase diagram is gathered by preparing heating or cooling curves for different pressures.

The **triple point** is the value of temperature and pressure at which there is equilibrium among the solid, liquid, and gas phases of a substance.

The **critical temperature (T_c)** is the temperature above which thermal motion is so violent that a gas cannot be liquefied; that is, the liquid state cannot exist above T_c for that substance.

The **critical pressure (P_c)** is the pressure that liquefies a gas at its critical temperature.

A **supercritical fluid** is a substance existing at a temperature above its critical temperature.

A **supercooled liquid** is a liquid at a temperature below its normal freezing point.

If a liquid is cooled well below its normal freezing point, it can solidify into a **glass**—a permanently supercooled liquid with an internal structure characteristic of the liquid—not the crystalline solid state. Glass is also known as an **amorphous solid**.

Theme 12 SOLUTIONS

*T*hus far we have been considering pure substances, either elements or compounds. We now turn to a discussion of homogeneous mixtures, otherwise known as solutions. Although there are many types of solutions and solvents, we will concentrate on water as the solvent because most solutions you come in contact with are aqueous solutions. Several examples of calculations involving solutions are given in Keys 53 and 55.

Key 50   The dissolving process

OVERVIEW   *The process of dissolving a solute in a given solvent is complex; it involves interactions among the various species present. In this Key, we will address the qualitative aspects of the dissolving process.*

The process of dissolving can be hypothetically broken down into at least three steps:

1.   Attractive forces of interaction between solute species in the pure solute must be overcome; this costs energy.
2.   Attractive forces of interaction between solvent species in a pure solvent must be overcome (i.e., a "hole" must be made in the solvent to admit solute species); this costs energy.
3.   The presence of solute species in the proximity of solvent species may result in attractive intermolecular interactions; this gives energy back.

For a substance to be soluble in a given solvent, the energy gained in step 3 must be greater than the sum of the energetic costs represented by steps 1 and 2. For this reason, it is very difficult to predict solubility.

Typically, ionic compounds dissolve with **dissociation**, the formation of ions in solution.

On the other hand, although molecular substances usually dissolve as molecules, they sometimes break up in polar solvents with the formation of ions. This process is sometimes called **ionization** when a compound separates into ions when dissolved in a polar solvent such as water.

A **strong electrolyte** is a solute that is essentially completely ionized in solution, giving a solution that is a good conductor of electricity.

A **weak electrolyte** is a solute that is only slightly ionized in solution, giving a solution that is a weak conductor of electricity.

A **nonelectrolyte** is a solute that dissolves to give a nonconducting solution.

The dissolving process is helped by **hydration**, the dipole-ion interaction in solutions between the solvent water molecules and the ions in solution.

An **unsaturated solution** is a solution that contains less than the amount of solute in a saturated solution at a particular temperature. It is therefore capable of dissolving additional solute.

A **saturated solution** is a solution that has dissolved solute in dynamic equilibrium with undissolved solute. It contains the maximum amount of dissolved solute at a particular temperature.

A **supersaturated solution** is an unstable solution that contains more solute than it would in the equilibrium state.

Key 51 Solubility

OVERVIEW *It is very difficult to predict the solubility of a compound because of the complexity of the interactions between solute and solvent species. However, qualitative rules of solubility have been developed.*

The **solubility** of a compound is the degree to which that compound dissolves in a solvent and is generally expressed either in moles of solute per liter of solution, molarity (Key 52), or in grams of solute per 100 g of solution.

A good general rule for predicting solubility is that **like tends to dissolve like**; that is, polar solvents tend to be good solvents for ionic compounds or polar molecular compounds, and nonpolar solvents tend to be good solvents for nonpolar molecular compounds.

A **hydrophilic** substance is attracted strongly to water molecules and therefore tends to be soluble in water.

A **hydrophobic** substance is attracted very weakly to water molecules and therefore tends to be insoluble in water.

Some general **solubility rules** have been developed:

1. All compounds of the alkali metals (Group IA) and the ammonium ion (NH_4^+) are soluble.
2. All compounds containing nitrate (NO_3^-), acetate ($C_2H_3O_2^-$), chlorate (ClO_3^-), and perchlorate (ClO_4^-) are soluble.
3. Most hydroxides (OH^-) are insoluble, except those of the alkali metals and $Ba(OH)_2$ and $Ca(OH)_2$.
4. Most compounds containing Cl^-, Br^-, and I^- are soluble, except those containing Ag^+, Hg_2^{2+}, and Pb^{2+}.
5. All carbonates (CO_3^{2-}), phosphates (PO_4^{3-}), and sulfides (S^{2-}) are insoluble, except those of the alkali metals and NH_4^+.
6. Most sulfates (SO_4^{2-}) are soluble. $Ca(SO)_4$ and $Ag_2(SO)_4$ are slightly soluble; $Ba(SO)_4$, $Hg(SO)_4$, and $Pb(SO)_4$ are insoluble.

The solubility of most compounds increases with increasing temperature, but a substance that is **retrograde soluble** is one whose solubility decreases with increasing temperature.

Henry's law states that the solubility of a gas in a liquid is directly proportional to the partial pressure of the gas over the resultant solution.

Key 52 Concentration expressions

OVERVIEW *Because chemistry is a quantitative science, it is necessary to define quantitative ways of expressing the concentration of the solute in a solution.*

Molarity (M) is concentration expressed as number of moles of solute per liter of solution:

$$M = \frac{\text{no. of moles of solute}}{1\,\text{L of solution}}$$

The molarity can be used to convert number of moles to volume (and vice versa). To determine the number of moles of solute in V liters of solution, multiply the molarity of the solution, M, by the number of liters, V:

$$M\,(\text{in mol/liter}) \times V\,(\text{in L}) = \text{no. of moles}$$

Molality (m) is concentration expressed as number of moles of solute per kilogram of solvent:

$$m = \frac{\text{no. of moles of solute}}{\text{kg of solvent}}$$

Since mass does not change with temperature, m is temperature-independent.

Mole fraction is the ratio of the number of moles of a component to the total number of moles present.

Mole percent equals mole fraction times 100%.

Mass percent is the mass of the solute divided by the total mass of the solution multiplied by 100%.

$$\text{mass \%} = \left(\frac{\text{mass of solute}}{\text{mass of solution}}\right) \times 100\%$$

The term **parts per million (ppm)** is used to measure the concentration of impurities in dilute solutions. 1 ppm = 1 part solute per 10^6 parts solution, by weight. In dilute aqueous solutions, 1 ppm = 1 mg/L.

As an example of the use of the unit ppm, if the concentration of sulfate ion (SO_4^{2-}) in a sample of water is found to be 375 ppm, calculate its molarity.

Recall that for dilute aqueous solutions, ppm is equivalent to mg/L. Therefore:

$$375 \text{ ppm } SO_4^{2-} = \frac{375 \times 10^{-3} \text{g } SO_4^{2-}}{\text{L solution}}$$

Since molarity has the units mol/L, g SO_4^{2-} must be converted to mol SO_4^{2-} using the formula weight:

$$\frac{350 \times 10^{-3} \text{ g } SO_4^{2-}}{\text{L solution}} \times \frac{\text{mol } SO_4^{2-}}{96.06 \text{ g } SO_4^{2-}} = \frac{0.336 \text{ mol } SO_4^{2-}}{\text{L solution}}$$

$$= 0.336 \, M \, SO_4^{2-}$$

Mass/volume percent is the mass of solute divided by the volume of solution times 100%.

$$\text{mass/vol } \% = \left(\frac{\text{mass of solute}}{\text{vol of solution}} \right) \times 100\%$$

Volume percent is the volume of solute divided by the volume of solution times 100%.

$$\text{vol } \% = \left(\frac{\text{vol of solute}}{\text{vol of solution}} \right) \times 100\%$$

Key 53 Concentration calculations

OVERVIEW *In this Key, examples of calculations involving solutions are presented.*

Molarity is used as a **conversion factor** to convert volume of solution to moles of solute.

KEY PROBLEM

To calculate the volume of 1.86 *M* Na_3PO_4 solution that can be prepared from 5.45 g Na_3PO_4, we first convert grams of Na_3PO_4 to moles of Na_3PO_4 present, using formula mass as a conversion factor, and then use molarity as a conversion factor to calculate the volume of Na_3PO_4 solution that can be prepared.

$$5.45 \text{ g} \times \frac{1 \text{ mol } Na_3PO_4}{163.94 \text{ g } Na_3PO_4} = 0.0332 \text{ mol } Na_3PO_4$$

$$0.0332 \text{ mol } Na_3PO_4 \times \frac{1 \text{ L}}{1.86 \text{ mol } Na_3PO_4} = 0.618 \text{ L}$$

To calculate the number of grams of $CoCl_2$ needed to prepare 75 mL of an 0.800 *M* $CoCl_2$ solution, we first must calculate the number of moles of $CoCl_2$ required:

$$\frac{0.800 \text{ mol } CoCl_2}{1 \text{ L}} \times 0.075 \text{ L} = 0.060 \text{ mol } CoCl_2$$

We then use the formula weight of $CoCl_2$ to calculate the required number of grams of $CoCl_2$:

$$0.060 \text{ mol } CoCl_2 \times \frac{129.83 \text{ g } CoCl_2}{1 \text{ mol } CoCl_2} = 7.8 \text{ g } CoCl_2$$

In order to convert molarity (a mass-volume unit) to molality (a mass-mass unit), the density of the solution must be known.

A solution containing 45.3 g NaCl per liter has a density of 1.020 g/cm³. To calculate *M*, grams of NaCl must be converted to moles of NaCl using the formula mass as a conversion factor. Then divide by the volume to obtain the molarity:

$$\frac{45.3 \text{ g NaCl}}{1 \text{ L}} \times \frac{1 \text{ mol NaCl}}{58.44 \text{ g NaCl}} = \frac{0.775 \text{ mol NaCl}}{1 \text{ L}} = 0.775 \, M \text{ NaCl}$$

To determine the molality of the solution, the volume of solution is converted to mass (using its density):

$$1000 \text{ mL solution} \times 1.020 \text{ g/cm}^3 = 1020. \text{ g solution}$$

The mass of solvent is obtained by subtracting grams of NaCl from grams of solution (1020. g – 45.3 g = 975 g solvent), and the molality is obtained as follows:

$$\frac{0.775 \text{ mol NaCl}}{0.975 \text{ kg solvent}} = 0.795 \, m$$

Dilution is the process of preparing a less concentrated solution from a more concentrated one.

To prepare 1.60 L of 0.800 M NaOH solution from a 1.40 M NaOH solution, first determine the number of moles of NaOH you need for the new solution:

$$\frac{0.800 \text{ mol NaCl}}{1 \text{ L}} \times 1.60 \text{ L} = 1.28 \text{ mol NaCl}$$

Then calculate the volume of the old solution needed to give you this number of moles:

$$1.28 \text{ mol NaCl} \times \frac{1 \text{ L}}{1.40 \text{ mol NaCl}} = 0.914 \text{ L}$$

To prepare the new solution, measure out exactly 0.914 L of 1.40 M NaOH solution and dilute to 1.60 L by adding pure water. Alternately, you can use the following formula:

$$M_I V_I = M_F V_F$$

Where M_I and M_F represent the initial and final concentrations, respectively, and V_I and V_F represent the initial and final volumes of the dilution process, respectively.

Key 54   Net ionic equations

OVERVIEW *When considering reactions occurring in solution, it is sometimes convenient to include only those ions that are actually involved in the reaction, or to write a net ionic equation describing the reaction.*

A **net ionic equation** includes only those species in solution that enter into the chemical reaction.

A **spectator ion** is an ion present in the solution but not participating in the chemical reaction.

Crystallization is the process by which dissolved solute precipitates from a solution and forms crystals.

The **precipitate** is the insoluble solid that forms in a solution.

Let us consider the reaction of a solution of sodium sulfate (Na_2SO_4) with a solution of barium nitrate [$Ba(NO_3)_2$]. On mixing these solutions, we observe a white solid precipitating. According to the solubility rules, this solid could be barium sulfate ($BaSO_4$). An equation describing this reaction is as follows:

$$Na_2SO_4(aq) + Ba(NO_3)_2(aq) \rightarrow BaSO_4(s) + 2\,NaNO_3(aq)$$

However, both sodium sulfate and barium nitrate, as well as the product sodium nitrate ($NaNO_3$), are dissociated in solution, resulting in the following equation:

$$2\,Na^+(aq) + SO_4^{2-}(aq) + Ba^{2+}(aq) + 2\,NO_3^-(aq) \rightarrow$$
$$BaSO_4(s) + 2\,Na^+(aq) + 2\,NO_3^-(aq)$$

Cancelling out all ions that appear on both sides of the equation gives us the net ionic equation:

$$Ba^{2+}(aq) + SO_4^{2-}(aq) \rightarrow BaSO_4(s)$$

Those ions that did not enter into the reaction (those cancelled out), in this case $Na^+(aq)$ and $SO_4^{2-}(aq)$, are the spectator ions.

Key 55 Solution stoichiometry

OVERVIEW *A large number of chemical reactions take place in solution. In this Key, we will consider stoichiometry in aqueous solutions.*

Titration is the reaction of an amount of a substance with an exact amount of a second substance until the complete reaction occurs.

A solution of $AgNO_3$ reacts with an $(NH_4)_2S$ solution to form Ag_2S according to the following reaction:

$$2\,AgNO_3(aq) + (NH_4)_2S(aq) \rightarrow Ag_2S(s) + 2\,NH_4NO_3(aq)$$

To calculate the amount of $0.500\,M$ $(NH_4)_2S$ solution needed to completely react with $0.375\,L$ of $0.600\,M$ $AgNO_3$ solution, first determine the number of moles of $AgNO_3$ in the solution:

$$\frac{0.600\text{ mol }AgNO_3}{1\text{ L}} \times 0.375\text{ L} = 0.225\text{ mol }AgNO_3$$

Then, using the molar ratio from the balanced equation, determine the number of moles of $(NH_4)_2S$ needed and convert this back to volume of $(NH_4)_2S$ solution:

$$0.225\text{ mol }AgNO_3 \times \frac{1\text{ mol }(NH_4)_2S}{2\text{ mol }AgNO_3} \times \frac{1\text{ L}}{0.500\text{ mol }(NH_4)_2S}$$

$$= 0.225\text{ L }(NH_4)_2S\text{ soln}$$

KEY PROBLEM 1

As another example of a solution stoichiometry problem, Cr^{3+} is quantitatively precipitated from solution as Cr_2S_3 by the addition of sulfide ion (S^{2-}). If all the Cr^{3+} is completely precipitated from 365 mL of a solution of $CrCl_3$ by the addition of 245 mL of $0.530\,M$ Na_2S solution, how many grams of $CrCl_3$ were present in the original solution?

First write the net ionic equation for the reaction:

$$2\,Cr^{3+}(aq) + 3\,S^{2-}(aq) \rightarrow Cr_2S_3(s)$$

Then determine the number of moles of S^{2-} that reacted in solution, noting that there is 1 mol of S^{2-} for every mole of Na_2S in solution:

$$\frac{0.530\text{ mol }S^{2-}}{1\text{ L}} \times 0.245\text{ L} = 0.130\text{ mol }S^{2-}$$

Then use the stoichiometry of the balanced equation to convert mol S^{2-} to mol Cr^{3+}:

$$0.130 \text{ mol } S^{2-} \times \frac{2 \text{ mol } Cr^{3+}}{3 \text{ mol } S^{2-}} = 0.0867 \text{ mol } Cr^{3+}$$

Since all this Cr^{3+} must have come from the $CrCl_3$ in the original solution, the molarity of that solution (after converting mL to L) is given by

$$\frac{0.08647 \text{ mol } CrCl_3}{0.365 \text{ L solution}} = 0.237 \text{ mol/L}$$

KEY PROBLEM 2

As a final example of a solution stoichiometry problem, calculate the number of grams of $Ni_3(PO_4)_2$ that will precipitate according to the following reaction:

$$3 \text{ Ni}^{2+}(aq) + 2 \text{ PO}_4^{3-}(aq) \rightarrow Ni_3(PO_4)_2(s)$$

If 475 mL of $0.400 M$ $NiCl_2$ solution is added to 245 mL of $0.400 M$ $(NH_4)_3PO_4$ solution, we first must determine the number of moles of Ni^{2+} and PO_4^{3-} in the original solutions:

$$\frac{0.400 \text{ mol } NiCl_2}{1 \text{ L}} \times 0.475 \text{ L} \times \frac{1 \text{ mol } Ni^{2+}}{\text{mol } NiCl_2} = 0.190 \text{ mol } Ni^{2+}$$

$$\frac{0.400 \text{ mol } (NH_4)_3PO_4}{1 \text{ L}} \times 0.245 \text{ L} \times \frac{1 \text{ mol } PO_4^{3-}}{\text{mol}(NH_4)_3PO_4} = 0.0980 \text{ mol } PO_4^{3-}$$

Since the stoichiometry of the reaction states that we need 2 mol PO_4^{3-} for every 3 mol Ni^{2+}, we do the calculation to see which reactant is the limiting reactant by asking how many moles of Ni^{2+} are needed to react with the given 0.0980 mol of PO_4^{3-}:

$$0.0980 \text{ mol } PO_4^{3-}\left(\frac{3 \text{ mol } Ni^{2+}}{2 \text{ mol } PO_4^{3-}}\right) = 0.147 \text{ mol } Ni^{2+}$$

We were given 0.190 mol Ni^{2+}, which is more than we need to react all the phosphate. Therefore the PO_4^{3-} is the limiting reactant and we calculate the amount of product from the given amount of PO_4^{3-}.

$$0.0980 \text{ mol } PO_4^{3-} \times \frac{1 \text{ mol } Ni_3(PO_4)_2}{2 \text{ mol } PO_4^{3-}} \times \frac{336.01 \text{ g } Ni_3(PO_4)_2}{\text{mol } Ni_3(PO_4)_2}$$

$$= 16.5 \text{ g } Ni_3(PO_4)_2$$

Key 56 Colligative properties

OVERVIEW *Colligative properties are properties of solutions that result from the presence of a solute but are independent of that specific solute. The four colligative properties are vapor pressure lowering, boiling point elevation, freezing point depression, and osmotic pressure.*

An **ideal solution** is a solution in which there is negligible interaction between solute molecules. Ideal solutions obey Raoult's law.

Vapor pressure lowering is the decrease in the vapor pressure of a solution below the vapor pressure of the pure solvent.

This definition is stated formally by **Raoult's law** which says that the presence of a solute lowers the vapor pressure of the pure solvent in proportion to the solute concentration.

Vapor pressure is defined as the pressure exerted by the vapor over a liquid.

Boiling point elevation is the increase in boiling point of a solution above the boiling point of the pure solvent. It is equal to the boiling point elevation constant (K_b, characteristic of the solvent) times the molality of the solution:

$$\Delta T_b = K_b \cdot m$$

Freezing point depression is the decrease in the freezing point of a solution below the freezing point of the pure solvent. It is equal to the freezing point depression constant (K_f, characteristic of the solvent) times the molality of the solution:

$$\Delta T_f = K_f \cdot m$$

Osmosis results from the passage of solvent molecules through a semi-permeable membrane into a solution of higher concentration.

Osmotic pressure is the external pressure necessary to prevent osmosis. Osmotic pressure is calculated as follows:

$$\Pi = MRT$$

A **semipermeable membrane** is a membrane through which only certain ions or molecules can move.

Isotonic solutions are solutions that have the same osmotic pressure.

Theme 13 CHEMICAL EQUILIBRIUM

*I*n Key 46, we introduced the concept of vapor pressure as an example of a physical equilibrium—an equilibrium between two different physical states. In this Theme, we will discuss chemical equilibria that exist between different substances under conditions of chemical reaction.

Key 57 The law of chemical equilibrium

OVERVIEW *The equilibrium expression is a quantitative statement of the condition of chemical equilibrium.*

In general, a state of **equilibrium** exists when there are no observable changes in the system with time.

A **dynamic equilibrium** exists in a system in which the rates of the forward and reverse reactions are equal.

A **physical equilibrium** is a dynamic equilibrium that involves a physical transformation.

A **chemical equilibrium** is a dynamic equilibrium that involves a chemical reaction.

The **law of chemical equilibrium** describes the relationships among the concentrations of reactants and products when a system is at chemical equilibrium. It consists of two parts, an equilibrium constant and the mass action expression.

An **equilibrium constant** (K) is an experimentally determined number that describes the relative concentrations of reactants and products.

The **mass action expression** for a chemical equilibrium

$$a\text{A} + b\text{B} + \cdots \rightleftarrows c\text{C} + d\text{D} + \cdots$$

is equal to the product of the concentrations of each product raised to a power equal to its coefficient in the balanced equation divided by the product of the concentrations of each reactant raised to a power equal to its coefficient in the balanced equation:

$$\frac{[\text{C}]^c [\text{D}]^d \cdots}{[\text{A}]^a [\text{B}]^b \cdots}$$

The **reaction quotient** (Q) is the numerical value of the mass action expression. If the system is at equilibrium, $Q = K$. If $Q > K$, the reaction will go in the reverse direction to reach equilibrium. If $Q < K$, the reaction will proceed in the forward direction to reach equilibrium.

At equilibrium, the numerical value of the mass action expression Q is equal to the equilibrium constant K. This is an expression of the law of chemical equilibrium. For the above general equation,

$$K = \frac{[\text{C}]^c [\text{D}]^d \cdots}{[\text{A}]^a [\text{B}]^b \cdots}$$

Key 58   The magnitude of the
equilibrium constant

OVERVIEW *The magnitude of the equilibrium constant, which is determined experimentally, is a measure of the position of equilibrium.*

The **numerical value** of the equilibrium constant must be experimentally determined and is dependent only on that particular equilibrium and on temperature.

As an example, PCl_5 decomposes into PCl_3 and Cl_2 in the gas phase according to the following equation:

$$PCl_5(g) \rightleftarrows PCl_3(g) + Cl_2(g)$$

2.00 mol PCl_5 is placed in a 1-L container at 546 K. At equilibrium, the concentrations of the components are found to be the following:

$$[PCl_3] = [Cl_2] = 0.347 \, M; [PCl_5] = 1.65 \, M$$

The value of K is found by inserting these equilibrium concentrations into the equilibrium expression as follows:

$$K = \frac{[PCl_3][Cl_2]}{[PCl_5]} = \frac{(0.347)(0.347)}{(1.65)} = 0.0730$$

The **magnitude** of the equilibrium constant is a measure of the position of the equilibrium, that is, whether it lies to the left (little reaction of reactants to form products) or to the right (relatively complete reaction of reactants).

As a rule of thumb, if the value of the equilibrium constant K is

1. very small (less than 10^{-5}), then the numerator of the mass action expression at equilibrium is small compared to the denominator (the product concentrations are much smaller than the reaction concentrations at equilibrium) and the forward reaction is unfavorable compared to the reverse reaction (the equilibrium lies to the left).

2. very large (greater than 10^5), then the product concentrations are much larger than the reaction concentrations at equilibrium and the forward reaction is favored compared to the reverse reaction (the equilibrium lies to the right).

3. neither very large nor very small, then the numerator and the denominator of the mass action expression are more balanced at equilibrium and there are significant concentrations of both reactants and products at equilibrium.

Key 59 Le Chatelier's principle

OVERVIEW *Although a system at equilibrium tends to remain that way, the equilibrium may be disturbed by a stress that causes the position of equilibrium to change.*

Le Chatelier's principle states that when a system at equilibrium is disturbed by a stress, the position of the equilibrium shifts to relieve that stress and to establish a new equilibrium condition.

There are several ways that an equilibrium involving gases can be disturbed:

1. by adding or removing a component
2. by changing the temperature
3. by changing the pressure.

As an example of **adding or removing a component**, consider the gas phase equilibrium between nitrogen (N_2) and hydrogen (H_2) forming ammonia (NH_3):

$$N_2(g) + 3 H_2(g) \rightleftarrows 2 NH_3(g)$$

If a stress is introduced by adding N_2 to the system at equilibrium, Le Chatelier's principle says that the system will act in a manner to reduce the amount of N_2; that is, the forward reaction will proceed until equilibrium is reestablished.

In terms of the law of chemical equilibrium, at equilibrium the mass action expression is equal to the equilibrium constant:

$$K = \frac{[NH_3]^2}{[N_2][H_2]^3}$$

When N_2 is added to the system, the mass action expression becomes smaller than K since its denominator increases. To reassume its equilibrium value, the numerator must increase and the denominator must decrease; that is, N_2 and H_2 must react to give more NH_3 (the reaction proceeds to the right). The same arguments hold for adding H_2 (shift to the right) or product NH_3 (shift to the left).

Raising the temperature of a system of an exothermic reaction at equilibrium (adding thermal energy) causes a shift to the left (the reverse endothermic reaction absorbs the added energy). In general,

- an increase in temperature favors an endothermic reaction.
- a decrease in temperature favors an exothermic reaction.

The **effect of changing pressure** depends on how that change occurs. If pressure is increased by decreasing the volume of the container, then an equilibrium involving a gas phase reaction will shift to the side with the fewest number of moles of species. In the above equilibrium, with 4 mol of gas on the left and 2 mol of gas on the right, the shift will be to the right. If pressure is raised by adding an inert gas (one not entering into the reaction), there will be no effect on equilibrium.

Key 60 Heterogeneous equilibria

OVERVIEW *The previous three Keys used gas phase equilibria as examples. Now we will turn to equilibria in systems involving two or more phases.*

A **heterogeneous equilibrium** is an equilibrium that involves two or more phases.

When writing the mass action expression for a heterogeneous equilibrium, only those concentrations that can be varied are included. We therefore omit the concentrations of all pure liquid and solid phases because they have fixed concentrations.

The **solubility product (K_{sp})** of a compound is an equilibrium constant that is a measure of the solubility of that compound.

As an example, the equilibrium describing dissolving $BaSO_4$, a relatively insoluble salt, is given by the following:

$$BaSO_4(s) \rightleftarrows Ba^{2+}(aq) + SO_4^{2-}(aq)$$

The corresponding equilibrium expression is as follows:

$$K_{sp} = [Ba^{2+}(aq)][SO_4^{2-}(aq)]$$

The concentration of $BaSO_4$ is not included on the right side of the equation since it is constant.

If it is found that a saturated solution of $BaSO_4$ contains $1.05 \times 10^{-5} M$ Ba^{2+} and $1.05 \times 10^{-5} M$ SO_4^{2-}, then the numerical value of K_{sp} is

$$K_{sp} = [Ba^{2+}][SO_4^{2-}] = (1.05 \times 10^{-5})(1.05 \times 10^{-5}) = 1.10 \times 10^{-10}$$

The **common ion effect** is a shift in equilibrium caused by the addition of a compound having an ion that is common to the dissolved species. This is an effect of Le Chatelier's principle on this type of equilibrium.

Continuing with the above example, if enough Na_2SO_4 (a soluble compound and a source of sulfate ion common to $BaSO_4$) is added to the saturated solution of $BaSO_4$ to bring the concentration of SO_4^{2-} to $1.00 \times 10^{-2} M$, then by rearranging the expression for K_{sp}, $[Ba^{2+}]$ can be calculated as follows:

$$[Ba^{2+}] = \frac{K_{sp}}{[SO_4^{2-}]} = \frac{1.10 \times 10^{-10}}{1.00 \times 10^{-2}} = 1.10 \times 10^{-8}$$

The resulting concentration is much lower than the concentration of barium ion in the saturated solution. This reduction in the concentration of Ba^{2+} is a result of the common ion effect.

Theme 14 ACIDS AND BASES

*A*cids and bases comprise important classes of compounds. The behavior of acids and bases and the reactions between them in aqueous solution provide important examples of chemical principles.

Key 61   Classification of acids and bases

OVERVIEW *There are several ways of defining or classifying acids and bases. The historical definition depends on observable phenomena, whereas modern concepts involve structure and chemical behavior.*

Historically, an **acid** was defined as a compound that tastes sour, feels prickly on the skin, turns the organic dye litmus red, and reacts with certain metals to produce hydrogen gas. In modern terms, an acid is a substance that, in aqueous solution, increases the concentration of H^+ (or H_3O^+; see below).

Similarly, a **base** was defined as a compound that tastes bitter, feels slippery, and turns litmus blue. We now consider a base to be a substance that, in aqueous solution, increases the concentration of OH^-.

The **hydronium ion (H_3O^+)** is the notation for a hydrated proton, that is, a proton in aqueous solution.

An **Arrhenius acid** is a compound that dissociates in aqueous solution to give H^+ (or H_3O^+) ions, whereas an **Arrhenius base** dissociates to give OH^- ions.

A **Brønsted-Lowry acid** is a species that donates a proton to another species in a reaction, whereas a **Brønsted-Lowry base** is the species that accepts a proton in that reaction.

A Brønsted acid (HX) and the base formed by the removal of a proton (X^-) are called **conjugate acid-base pairs**. X^- is the conjugate base of HX.

A **Lewis acid** is a species that accepts a pair of electrons from another species in a reaction, whereas a **Lewis base** is the species that donates a pair of electrons.

A Lewis acid can react with a Lewis base to form a **Lewis acid-base adduct** in which the base provides the pair of electrons forming the bond. For example, ammonia (a Lewis base) reacts with boron trifluoride (a Lewis acid):

$$NH_3 + BF_3 \rightarrow H_3N\!-\!BF_3$$

A **strong acid** is an acid that is essentially completely dissociated in solution (i.e., a strong electrolyte; Key 50). Common strong acids are hydrochloric (HCl), nitric (HNO_3), sulfuric (H_2SO_4), and perchloric ($HClO_4$) acids.

A **weak acid** is one that is only partially dissociated in solution (i.e., a weak electrolyte; Key 50).

An **amphoteric** (or **emphiprotic**) substance can behave as either an acid or a base.

An **acid anhydride** is a nonmetal oxide that reacts with water to form an acid.

A **base anhydride** is a metal oxide that reacts with water to form a base.

Hydrolysis is the reaction of a compound with water, usually resulting in a change in the acidity of the solution.

Key 62   Equilibria involving acids
and bases

OVERVIEW *Following the law of chemical equilibrium, one can write equilibrium expressions for acids or bases in aqueous solution.*

The **equilibrium constant for an acid** (K_a) is obtained by writing the law of chemical equilibrium for the dissociation of that acid in aqueous solution:

$$HX(aq) \rightleftarrows H^+(aq) + X^-(aq)$$

The equilibrium expression defining K_a is

$$K_a = \frac{[H^+][X^-]}{[HX(aq)]}$$

Alternatively, if the solvent (water) is included in the equilibrium (in which it accepts a proton and forms the hydronium ion, H_3O^+), then

$$HX(aq) + H_2O \rightleftarrows H_3O^+(aq) + X^-(aq)$$

The corresponding equilibrium expression is

$$K_a = \frac{[H_3O^+][X^-]}{[HX(aq)]}$$

Note that the concentration of water is not included in the equilibrium expression since water is a pure substance whose concentration does not vary.

In like manner, the **equilibrium constant for a base** (K_b) corresponds to the dissociation of a base, M, in aqueous solution:

$$M(aq) + H_2O \rightleftarrows MH^+(aq) + OH^-(aq)$$

$$K_b = \frac{[MH^+][OH^-]}{[M]}$$

Neutral water self-dissociates (sometimes called **autoionization**) according to the following equation:

$$H_2O + H_2O \rightleftarrows H_3O^+(aq) + OH^-(aq)$$

or, leaving out the solvent species,

$$H_2O \rightleftarrows H^+(aq) + OH^-(aq)$$

The **equilibrium constant for water (K_w)** is then given by the following expression:

$$K_w = [H^+][OH^-] = 1.0 \times 10^{-14}$$

This experimentally determined value (at 25°C) means that in pure water $[H^+] = 1 \times 10^{-7}$ (since $[H^+] = [OH^-]$; $[H^+]^2 = 1.0 \times 10^{-14}$, and $[H^+] = 1.0 \times 10^{-7}$).

Key 63   Acid strength

OVERVIEW *Using the Brønsted-Lowry concept of acids and bases, one can list acids in order of their strength. This order is important in predicting the direction in which an acid-base reaction will proceed.*

According to the **Brønsted-Lowry** definition of an acid (Key 61), the reaction of the acid HX with water, which acts as a base by accepting a proton, is written as follows:

$$HX(aq) + H_2O \rightarrow H_3O^+(aq) + X^-(aq)$$

A **table of relative acid-base strengths** can be constructed by listing acids in order of the ease with which they transfer a proton to the solvent water.

K_a	Acid					Base	K_b
very large	$HClO_4$	+	H_2O	$\rightleftarrows$	H_3O^+ +	ClO_4^-	very small
very large	HCl	+	H_2O	$\rightleftarrows$	H_3O^+ +	Cl^-	very small
	H_3O^+	+	H_2O	$\rightleftarrows$	H_3O^+ +	H_2O	
6.9×10^{-4}	HF	+	H_2O	$\rightleftarrows$	H_3O^+ +	F^-	1.4×10^{-11}
1.8×10^{-5}	$HOAc$	+	H_2O	$\rightleftarrows$	H_3O^+ +	OAc^-	5.6×10^{-10}
5.6×10^{-10}	NH_4^+	+	H_2O	$\rightleftarrows$	H_3O^+ +	NH_3	1.8×10^{-5}
very small	H_2O	+	H_2O	$\rightleftarrows$	H_3O^+ +	OH^-	very large

The **strongest acid** is at the upper left corner of the table because it has the greatest tendency to donate a proton to water. The **conjugate base of the strongest acid is the weakest base** (at the upper right corner of the table).

The **weakest acid** is at the lower left corner of the table because it has the least tendency to donate a proton to water. The **conjugate base of the weakest acid is the strongest base** (at the lower right corner of the table).

The **relative magnitudes of K_a and K_b** have been included in the table. K_a decreases in going down the table (left side). K_b increases in going up the table (for bases, from the lower right to upper right of the table).

An acid **reacts** with a base whose conjugate acid lies below it in the table because the acid formed will be weaker (have less of a tendency to dissociate).

The **leveling effect** results from the inability of a solvent to differentiate among the relative strengths of acids stronger than the solvent's conjugate acid.

Key 64 Equilibrium calculations

OVERVIEW *In this Key, we will illustrate simple calculations involving solutions of a strong and a weak acid.*

To calculate the H^+ concentration in a $0.010\,M$ solution of HCl, realize that HCl is a strong acid that is completely dissociated in solution:

$$HCl(aq) \rightleftarrows H^+(aq) + Cl^-(aq)$$

Therefore, a $0.010\,M$ solution of HCl contains $0.010\,M$ of H^+. As an example of a weak acid, we will consider a $0.010\,M$ solution of acetic acid (abbreviated HOAc) that has $K_a = 1.8 \times 10^{-5}$. In solution,

$$HOAc \rightleftarrows H^+ + OAc^-$$

OAc^- is the abbreviation for the acetate ion. The equilibrium expression is

$$K_a = \frac{[H^+][OAc^-]}{[HOAc]}$$

To solve this problem, realize that every molecule of HOAc that dissociates gives one H^+ and one OAc^-. Furthermore, if $[H^+]$ is said to be equal to x, then at equilibrium, $[H^+] = [OAc^-] = x$ and $[HOAc] = (0.010\,M - x)$. Plugging these numbers and the value for K_a into the above equilibrium expression and solving for x gives the following:

$$1.8 + 10^{-5} = \frac{[x][x]}{(1.8 + 10^{-5})}$$

$$x^2 = (1.8 + 10^{-5})(0.010 - x)$$

Since HOAc is a weak acid, we can assume that $x \ll 0.010$, and this reduces to the following:

$$x^2 = (1.8 + 10^{-5})(0.010)$$

$$x^2 = 3.16 \times 10^{-4} = [H^+]$$

The assumption is verified since the calculated value of x, 0.000316, is much less than 0.010.

Key 65 The pH scale

OVERVIEW *Because the concentrations of H^+ and OH^- can vary over many orders of magnitude (10^{-14} to 10^{14}), a logarithmic scale is defined to express these concentrations.*

In general, a **pX scale** is a way to express the value of X as a logarithm. For example, if X represents a concentration ($[X]$), then $pX = -\log[X]$.

pH is another way of expressing the $[H_3O^+]$ of a solution. It is therefore a measure of the acidity of a solution; $pH = -\log[H_3O^+]$.

pOH is a measure of the basicity $[OH^-]$ of a solution; $pOH = -\log[OH^-]$.

In a **neutral solution**, $[H_3O^+] = [OH^-] = 1.0 \times 10^{-7}$. Therefore the pH of a neutral solution is 7.00 (at 25°C).

In **acidic solutions**, the pH < 7.00. For example, a solution in which $[H_3O^+] = 10^{-3}$, the pH = 3.00.

In **basic solutions**, the pH > 7.00. For example, a solution in which $[OH^-] = 10^{-4}$, the pH = 10.00 because this corresponds to $[H_3O^+] = 10^{-10}$.

By taking the negative logarithm of each side of the expression for K_w (Key 62),

$$[H^+][OH^-] = 1.0 \times 10^{-14}$$

and

pH + pOH = 14.00

Key 66   Reactions of acids and bases

OVERVIEW *Acids and bases react with one another in stoichiometric amounts to produce a neutral solution.*

A **neutralization** reaction occurs when an acid and a base are mixed in such proportions that the resulting solution has neither acidic nor basic properties.

Equimolar mixtures of acids and bases are therefore said to **neutralize** one another.

The result of a neutralization reaction in aqueous solution is the production of solvent molecules and a **salt**, a neutral compound that is neither acidic nor basic. As an example of a neutralization reaction, nitric acid reacts with sodium hydroxide (a base) to produce water and the salt sodium nitrate ($NaNO_3$):

$$HNO_3(aq) + NaOH(aq) \rightarrow H_2O(aq) + NaNO_3(aq)$$

The process of **titration** is the determination of the amount of a substance by adding known amounts of a second substance until a complete reaction occurs.

For acids and bases, an **equivalent** is that amount of substance that produces or reacts with one mole of H^+.

The **endpoint** (or **equivalence point**) is the point in a titration when equal numbers of equivalents of reactants have been combined; that is, the completion of the reaction.

An **indicator** is a compound (usually a weak acid) that undergoes a color change with a change in pH. That is, the weak acid and its conjugate base have different colors. Indicators are used to determine the endpoints of titrations.

A **binary acid** is a substance with the general formula H_nX that produces an acidic aqueous solution.

A **monoprotic acid** is an acid that can furnish one H^+ per molecule.

A **diprotic acid** is an acid that can furnish two H^+ per molecule.

An **oxyacid** is an oxygen-containing acid, such as nitric acid (HNO_3) and sulfuric acid (H_2SO_4).

Key 67 Titration reactions

OVERVIEW *In this Key, we will illustrate, using net ionic equations (Key 54), typical titration reactions between strong and weak acids and bases.*

The reaction of a **strong acid and a strong base** is just the reaction between H_3O^+ and OH^- because both are completely dissociated in solution. The neutralization reaction between hydrochloric acid and sodium chloride (in aqueous solution) is

$$H_3O^+(aq) + Cl^-(aq) + Na^+(aq) + OH^-(aq) \rightarrow$$
$$2H_2O + Na^+(aq) + Cl^-(aq)$$

Or, in net ionic form:

$$H_3O^+(aq) + OH^-(aq) \rightarrow 2H_2O$$

For the case of the reaction of a **strong acid and a weak base** one must realize that the weak base (NH_3, for example) is only slightly dissociated in solution. The neutralization reaction for the reaction of HCl and NH_3 is written as

$$H_3O^+(aq) + Cl^-(aq) + NH_3(aq) \rightarrow NH_4^+(aq) + Cl^-(aq) + H_2O$$

The net ionic equation is

$$H_3O^+(aq) + NH_3(aq) \rightarrow NH_4^+(aq) + H_2O$$

Similarly, for the reaction of a **weak acid and a strong base**, the acid is only slightly dissociated in solution. The neutralization reaction, using acetic acid and sodium hydroxide as examples, is

$$HOAc(aq) + Na^+(aq) + OH^-(aq) \rightarrow OAc^-(aq) + Na^+(aq) + H_2O$$

The net ionic equation is

$$HOAc(aq) + OH^-(aq) \rightarrow OAc^-(aq) + H_2O$$

In the reaction of a **weak acid and a weak base**, both are only slightly dissociated in solution. The net ionic equation for the neutralization reaction of acetic acid and ammonia is

$$HOAc(aq) + NH_3(aq) \rightarrow NH_4^+(aq) + OAc^-(aq)$$

Key 68 Buffer solutions

OVERVIEW *Solutions that contain similar amounts of a weak acid and a salt of its conjugate base are known as buffer solutions and are resistant to large changes in pH.*

A **buffer solution** is a solution made from a weak acid and a salt whose anion is the conjugate base of that acid (or a weak base and the conjugate acid of that base). Buffer solutions are resistant to large changes in pH.

To understand a buffer, consider a solution that contains similar amounts of acetic acid (a weak acid; Key 62) and sodium acetate, which contains the conjugate base of acetic acid, the acetate ion OAc⁻. The equilibrium that must be obeyed is for the dilute acid HOAc:

$$HOAc \rightleftarrows H^+ + OAc^-$$

The equilibrium expression is

$$K_a = \frac{[H^+][OAc^-]}{[HOAc]}$$

Solving this equation for [H⁺] gives

$$[H^+] = K_a \frac{[HOAc]}{[OAc^-]}$$

This equation states that [H+] is a function of K_a times the ratio of [HOAc] to [OAc⁻].

Taking the logarithm of each side and multiplying through by −1 gives, for the weak acid HA and its conjugate base A⁻,

$$pH = pK_a + \log \frac{[A^-]}{[HA]}$$

This equation is known as the **Henderson-Hasselbach equation** and is used to calculate the pH of a buffer.

Theme 15 THERMODYNAMICS

*T*hermodynamics is the study of energy changes and the flow of energy. The laws of thermodynamics not only govern energy changes in reactions but also provide a tool that allows the prediction of whether or not a given reaction or process will proceed spontaneously.

Key 69   The first law of thermodynamics

OVERVIEW *In Key 22, enthalpy changes that may occur during chemical and physical processes were discussed. In this Key, we will expand on those concepts.*

Formally, when discussing **conservation of energy** (Key 11), it is necessary to distinguish between a system and its surroundings.

When studying the flow of energy, the **system** is defined as that part of the universe that is being studied.

The **surroundings** are then defined as everything else in the universe.

A **closed system** is a system that allows the exchange of energy, but not mass, with its surroundings.

The **first law of thermodynamics** states that when a system undergoes a series of changes that eventually bring it back to its original state, the net energy change is zero; formally, $\Delta E = q + w$, where ΔE is the change in internal energy, q is the heat added to the system, and w is the work done on the system. This is just a restatement of the law of conservation of energy (Key 11).

The **internal energy** (*E*) is the total kinetic and potential energy of a system.

A **state function** is a quantity whose value depends only on the current state of a system. A change in its value depends only on the initial and final states and is independent of path. *E* and *H*, enthalpy (Key 22), are examples of state functions.

To measure energy changes during a reaction, one uses a **calorimeter** (Key 22). In an ordinary calorimeter, the reaction takes place at constant pressure. Under these conditions, the heat flow measured is equal to ΔH for the reaction, or in equation form, $\Delta H = q_{reaction}$ (constant pressure).

A **bomb calorimeter** is one in which the reaction takes place at constant volume; the heats measured are equal to ΔE for the reaction. In equation form, $\Delta E = q_{reaction}$ (constant volume).

The relationship between *H* and *E* is given as $\boldsymbol{\Delta H = \Delta E + \Delta(PV)}$. At constant pressure, this reduces to $\boldsymbol{\Delta H = \Delta E + P\Delta V}$, where $P\Delta V$ is an expression of the work done against the prevailing pressure.

Key 70   Entropy

OVERVIEW *The concept of entropy is a measure of the randomness or disorder of a system. The second and third laws of thermodynamics deal with entropy.*

The **second law of thermodynamics** states that when a spontaneous change occurs, it is accompanied by an increase in the total entropy of the universe. This is sometimes written as follows:

$$\Delta S_{universe} = \Delta S_{system} + \Delta S_{surroundings} > 0$$

Entropy (S) is a measure of the disorder of a system. For example, the entropy of one mole of a substance in the gas phase is greater than in the liquid phase because a gas is much more disordered than a liquid. Similarly, the entropy of a substance in the liquid phase is greater than in the solid phase, and the entropy of an amount of a substance dissolved in solution is greater than in the pure substance.

The **third law of thermodynamics** states that the entropy of a pure crystalline substance at absolute zero is zero; $S = 0$ at $0\,K$.

Because one can measure the absolute entropy for a pure substance, the **standard entropy ($S°$)** of a substance is defined as the entropy of one mole of that substance at 25°C and 1 atm.

The **standard entropy change ($\Delta S°$)** for a reaction or physical process is defined as follows:

$$\Delta S° = \sum [(S°_{products})(mol_{products})] - \sum [(S°_{reactants})(mol_{reactants})]$$

Key 71 Free energy and spontaneity

OVERVIEW *The concept of change in free energy (which balances change in enthalpy with change in entropy) allows one to predict whether or not a reaction or process will proceed spontaneously.*

The **Gibbs free energy** (G) is the thermodynamic function relating the two quantities enthalpy (H) and entropy (S): $G = H - TS$, or for a process at constant T and P:

$$\Delta G = \Delta H - T\Delta S$$

The free energy function is the **criterion for spontaneity** because if ΔG is negative, then ($\Delta H - T\Delta S$) is negative and the process is spontaneous at constant T and P. Summarizing:

1. If ΔG is negative, the reaction is spontaneous.
2. If ΔG is positive, the reaction is not spontaneous.
3. If ΔG equals zero, the system is at equilibrium.

The **standard free energy change** ($\Delta G°$), that is, the free energy change at standard conditions of a pressure of 1 atm and temperature of 25°C, is defined as follows:

$$\Delta G° = \Delta H - T\Delta S°$$

For a reaction, $\Delta G° = (\Sigma \Delta G_f° \text{ products}) - (\Sigma \Delta G_f° \text{ reactants})$.

The **standard free energy change of formation** ($\Delta G_f°$) is the change in free energy when one mole of a compound in its standard state is formed from its elements in their standard states.

Key 72 Free energy and temperature

OVERVIEW *Using the concepts developed in the previous Key, we are now able to predict the effects of temperature on free energy change and reaction spontaneity.*

At the end of the last Key, the relationship among free energy, enthalpy, and entropy was given:

$$\Delta G° = \Delta H° - T\Delta S°$$

The form of the equation indicates that there is a balance between enthalpy and entropy to determine whether $\Delta G°$ is positive or negative. Although both $\Delta H°$ and $\Delta S°$ can be positive or negative, the deciding factor is the **effect of temperature** because the entropy term is multiplied by the absolute temperature. This is summarized in the following table.

Effect of the Numerical Sign of $\Delta H°$ and $\Delta S°$ on $\Delta G°$ and Spontaneity

Sign of $\Delta H°$	Sign of $\Delta S°$	Sign of $\Delta G°$	Comment
positive	positive	positive (low T)	heating makes reaction
		negative (high T)	spontaneous
negative	negative	negative (low T)	cooling makes reaction
		positive (high T)	spontaneous
negative	positive	negative	always spontaneous
positive	negative	positive	never spontaneous

Key 73   Free energy and equilibrium

OVERVIEW *In this Key, a relationship between change in free energy and the equilibrium constant will be developed, and it will be shown that the sign of $\Delta G°$ can be used to predict the magnitude of K.*

For a chemical reaction, the relationship between the free energy change under standard conditions ($\Delta G°$) and the free energy change under any conditions (ΔG) is given by the following expression:

$$\Delta G = \Delta G° + RT \ln Q$$

R is the gas constant (8.314 J/mol-K), T is the absolute temperature, and Q is the reaction quotient and is equal to the numerical value of the mass action expression.

For a **system at equilibrium**, $\Delta G = 0$ and $Q = K$. The above equation then becomes

$$0 = \Delta G° + RT \ln K$$

or, by rearranging

$$\Delta G° = -RT \ln K$$

Because the temperature is always a positive number, there are only the following three possibilities:

- $\Delta G°$ is negative. $K > 1$
- $\Delta G°$ equals 0. $K = 1$
- $\Delta G°$ is positive. $K < 1$

This is equivalent to the conclusions in the discussion of the magnitude of K in relation to the position of equilibrium (Key 58).

Theme 16 CHEMICAL KINETICS

*O*ur discussion of equilibrium and thermodynamics indicated that one can talk about the extent to which a reaction goes to completion, but nothing was said about how fast a reaction proceeds. The study of the rates of chemical reactions and the factors that influence them is known as chemical kinetics.

Key 74 The rates of chemical reactions

OVERVIEW *The rate at which a chemical reaction proceeds is affected by several factors, including concentration of reactants, temperature, and the presence of a catalyst. In this Key, we will introduce the concept of the rate equation.*

In general, when one speaks of the **rate** of a process, one means the speed at which the process proceeds.

The **reaction rate** of a chemical reaction is the rate of decrease of the concentration of a reactant or the rate of increase of the concentration of a product.

The **average rate** is obtained by measuring the concentration of a reactant or product at two different times, t_1 and t_2. The rate is equal to

$$\text{rate} = \frac{C_2 - C_1}{t_2 - t_1} = \frac{\Delta C}{\Delta t}$$

The units of the reaction rate are $\text{mol}\,\text{L}^{-1}\,\text{s}^{-1}$.

Because the rate of a reaction decreases as the reaction proceeds (the concentrations of reactants are decreasing), the **instantaneous rate**, or the reaction rate at a particular time, is obtained by taking the tangent to a plot of [C] versus time.

A **rate law** (or **rate equation**) is an equation that relates the rate of a reaction to the concentration of the reactants raised to the appropriate power. For example, for a reaction involving the species A and B, a general rate law is of the following form:

$$\text{rate} = k[\text{A}]^m[\text{B}]^n$$

The form of the rate equation (as well as the reaction order; see below) for a particular reaction must be determined experimentally.

The **rate constant** or the specific rate (k) is the proportionality constant in the rate law for a reaction. The dimensions for k depend on the overall reaction order.

The **reaction order** for each reactant is equal to the exponent in the rate equation for that reactant (m and n in the above rate equation).

The **overall reaction order** is the sum of the exponents in the rate law describing that reaction ($m + n$ in the above rate equation).

Key 75 Time dependence of
reactant concentrations

OVERVIEW *Rate equations give us information about the rate of a reaction as a function of reactant concentrations. These equations can be converted to equations that allow us to determine the concentration of reactants or products at any time during the reaction.*

The **integrated form of the rate equation** depends on the overall order of the particular reaction and allows one to determine the effect of time on reactant concentrations.

For a **first-order reaction**, where the rate equation is rate = $k[A]$, the relationship between concentration and time is as follows:

$$\ln[A] = -kt + \ln[A_0]$$

ln is the abbreviation for the natural logarithm, $[A_0]$ is the initial concentration (at $t = 0$) of A, and $[A]$ is the concentration of A at time t. Because this equation is in the form $y = mx + b$, the equation for a straight line, if a reaction is first-order, a plot of $\ln[A]$ versus t will give a straight line with slope (m) $-k$ and y-intercept $\ln[A_0]$.

The **half-life** $(t_{1/2})$ of a reaction is the time required for the concentration of a reactant to be reduced to one-half its initial value.

The **half-life for a first-order reaction**, which is independent of $[A_0]$, is given by

$$t_{1/2} = \frac{0.693}{k}$$

For a **second-order reaction**, where the rate equation is rate = $k[A]^2$, the relationship between concentration and time is as follows:

$$\frac{1}{[A]} = kt + \frac{1}{[A_0]}$$

Because this equation is also in the form of a straight line, if a reaction is second-order, a plot of $\frac{1}{[A]}$ versus t gives a straight line with slope k and y-intercept $\frac{1}{[A_0]}$.

The **half-life for a second-order reaction** is given by $t_{1/2} = \dfrac{1}{k[A_0]}$.
(Note that $t_{1/2}$ is independent of concentration but dependent on $[A_0]$.)

In a **zero-order reaction**, the rate equation is rate $= k[A]^0 = k$; that is, it is independent of concentrations. The relationship between concentration and time is as follows:

$$[A] = -kt + [A_0]$$

A plot of $[A]$ versus t gives a straight line with slope $-k$ and y-intercept $[A_0]$.

The **half-life for a zero-order reaction** is given by $t_{1/2} = \dfrac{[A_0]}{2k}$.

Key 76   Temperature dependence
of reaction rate

OVERVIEW *In order to describe the effect of temperature on the rates of chemical reactions, we must consider the mechanism by which simple reactions take place.*

The **reaction coordinate** is the path followed as reactant molecules approach each other, collide, and make and break chemical bonds, and the product species move apart.

The **activation energy** (E_a) is the minimum energy that must be surpassed in order for a chemical reaction to take place. Alternatively, it can be considered as the potential energy barrier to the reaction.

The **potential energy diagram** is a plot of potential energy versus reaction coordinate for a chemical reaction.

The **activated complex** or **transition state** is that intermediate species found during a chemical reaction when reactants have collided and are at the high point on the potential energy diagram for the reaction.

The distribution of kinetic energies in gases undergoing reaction obeys a **Boltzmann distribution** (Key 42).

The **Arrhenius equation** is a nonlinear equation relating the rate constant for a reaction k to the absolute temperature T:

$$k = Ae^{-(E_a/RT)}$$

A is a constant called the **frequency factor** (related to the frequency of each collision), E_a is the activation energy, and R is the gas constant (expressed as 8.314 J/mol-K).

Taking the natural logarithm of both sides of the Arrhenius equation gives the following:

$$\ln k = -\frac{E_a}{RT} + \ln A$$

This equation is in the form of a straight line. Therefore, a plot of k versus $1/T$ has a slope of $-E_a/R$ and a y-intercept of A.

Alternatively, **to determine E_a**, consider a reaction at two different temperatures (T_1 and T_2) with rate constants k_1 and k_2. Combining the two Arrhenius equations, simplifying, and rearranging gives the following:

$$\ln \frac{k_1}{k_2} = \frac{E_a}{R} \left(\frac{1}{T_2} - \frac{1}{T_1} \right)$$

KEY PROBLEM

If one knows the value of E_a and the rate constant at one temperature, this equation can also be used to calculate the rate constant at any other temperature. As an example, the gas-phase decomposition of HI into H_2 and I_2 is found to have an activation energy of 182 kJ/mol. At 700°C the rate constant k has a value of 1.57×10^{-3} M/s. Calculate the value of k at 850°C.

Using the equation above, insert the values of k_1, T_1, and T_2 (temperatures are always in K):

$$\ln \frac{k_1}{k_2} = \frac{E_a}{R} \left(\frac{1}{T_2} - \frac{1}{T_1} \right)$$

$$\ln \frac{(1.57 \times 10^{-3} \, M/s)}{k_2} = \frac{182 \times 10^3 \, J/mol}{8.314 \, J/mol \, K} \left(\frac{1}{1123 \, K} - \frac{1}{973 \, K} \right) = -3.005$$

Taking the antilog of both sides gives

$$\frac{1.57 \times 10^{-3} \, M/s}{k_2} = 4.95 \times 10^{-2}$$

$$k_2 = 3.17 \times 10^{-2} \, M/s$$

Key 77   Reaction mechanism

OVERVIEW *The mechanism of a reaction is a postulated series of steps that when summed produce the overall balanced equation for that reaction. Because it is theoretical in nature, any proposed reaction mechanism must be proven by experiment.*

A **reaction mechanism** provides a detailed model of how a reaction proceeds and includes the sequence of steps occurring during the process.

Collision theory is a theory of reaction rates that states that the rate of a reaction is proportional to the number of collisions per second between reactant molecules.

An **elementary process** is one of the individual steps in a multistep reaction mechanism.

The **molecularity** of a reaction specifies the number of molecules interacting in a single mechanistic step.

A **unimolecular reaction** is a step in which one molecule undergoes reaction.

A **bimolecular reaction** or **collision** is a reaction or collision between two molecules.

A **termolecular collision** is a collision involving three molecules in a single step.

The **rate-determining step** for a reaction is the slow step in a reaction mechanism that determines the rate at which products are formed.

The **molecularity of the rate-determining step** is reflected in the overall order of the rate equation.

A **catalyst** is a substance that speeds up or slows down a chemical reaction without being consumed. Typically, a catalyst acts by lowering the activation energy for that reaction.

A **chain reaction** is a reaction in which the product of one step is the reactant in another.

Theme 17 OXIDATION-REDUCTION REACTIONS AND ELECTROCHEMISTRY

*A*n important class of practical chemical reactions is known as oxidation-reduction (or redox) reactions. These reactions involve the transfer of one or more electrons between the reacting species. By carrying out these reactions under controlled conditions, one is able to convert chemical energy stored in the reactants to useful electrical energy.

INDIVIDUAL KEYS IN THIS THEME

Key 78 Oxidation states

OVERVIEW *The assignment of an oxidation state (or oxidation number) to an atom is a method designed to keep track of the electrons in species undergoing redox reactions.*

An **oxidation state**, or **oxidation number**, is assigned to each element in a compound or ion by assuming that the compound is an ionic substance. Some **heirarchical rules** are listed below.

1. Oxidation numbers must all add up to the charge on the substance.
2. Metals in groups IA and IIA and aluminum are +1, +2, and +3, respectively.
3. Hydrogen is +1, and fluorine is −1.
4. Oxygen is −2.
5. Group VII elements are −1.
6. Group VI elements are −2.
7. Group V elements are −3.

These rules are heirarchical, meaning that rule 1 takes precedence over all the others, rule 2 takes precedence over all but rule 1, and so on. In addition, considering two elements in the same group, the element closest to fluorine in the periodic table follows the rule.

For example: Determine the oxidation numbers in Na_2MoO_4. First we write the ions $2\ Na^+$ and MoO_4^{2-}. Sodium is +1, and we determine the oxidation number of Mo by constructing an equation as per rule 1.

$$\text{Charge} = (\text{ox no. of Mo}) + 4 \times (\text{ox no. of O})$$

We enter −2 for the charge and −2 for the oxidation number of oxygen to get

$$-2 = (\text{ox no. of Mo}) + 4 \times (-2)$$

When this equation is solved, we find that Mo has an oxidation number of +6, and we had previously assigned $Na = +1$ and $O = -2$.

Key 79 Oxidation-reduction reactions

OVERVIEW *An oxidation-reduction reaction is a chemical reaction in which there is a transfer of electrons from one chemical species to another, resulting in a change in oxidation state (electrons are negatively charged).*

Oxidation is the process of losing electrons, thereby increasing the species' oxidation number.

Reduction is the process of gaining electrons, thereby decreasing the species' oxidation number.

The **reducing agent** is the species that undergoes oxidation. (It provides the electrons necessary to reduce the other species and, in doing so, loses electrons, increases in positive charge, and goes to a higher oxidation state.)

The **oxidizing agent** is the species that undergoes reduction (it consumes the electrons lost by the species undergoing oxidation, decreases in positive charge, and goes to a lower oxidation state).

A simple redox reaction, in which metallic zinc reacts by losing two electrons, forming Zn^{2+} ions in solution, and Cu^{2+} ions gain two electrons to form metallic copper, can be represented by the following equation:

$$Zn(s) + Cu^{2+}(aq) \rightarrow Zn^{2+}(aq) + Cu(s)$$

The zinc is **oxidized** since its oxidation state changes from 0 to +2.

The Cu^{2+} ions are **reduced** since the copper changes from +2 to 0.

Every oxidation-reduction reaction can be separated into **half-reactions**—one involving the oxidation process, the other involving the reduction process.

Using the reaction above as an example, the **oxidation half-reaction** is

$$Zn(s) \rightarrow Zn^{2+}(aq) + 2e^-$$

and the **reduction half-reaction** is

$$Cu^{2+}(aq) + 2e^- \rightarrow Cu(s)$$

Disproportionation is the process in which the same chemical substance undergoes both oxidation and reduction.

Key 80   Balancing oxidation-
reduction equations

OVERVIEW *Many reactions can be balanced by inspection. Redox reactions often involve complex stoichiometry and large coefficients that are not obvious. The ion-electron method of balancing redox equations is the most reliable method for obtaining a valid balanced equation.*

The **ion-electron method** for balancing reactions takes more time but reliably produces the correct answer if attention is paid to all details. To balance a redox reaction:

1. Identify the two substances involved in the oxidation process and write them as one half-reaction. Identify the substances involved in the reduction process and use them to start the reduction half-reaction. (In fact, you do not need to know which is reduction and which is oxidation, but you do need two half-reactions.)

2. Balance all the atoms in the half-reactions except hydrogen and oxygen.

3. Balance the oxygen atoms by adding one water molecule for each oxygen needed.

4. Balance the hydrogen atoms by adding hydrogen ions (H^+).

5. Balance the charge by adding negatively charged electrons to the more positive side of the equations.

6. Equalize the number of electrons in the two half-reactions by multiplying the complete half-reactions by the appropriate small whole numbers. (Multiplying each half-reaction by the number of electrons in the other half-reaction always works.)

7. Add the equations, cancel the electrons (they *must* cancel), and cancel other like items such as H_2O and H^+ where possible. Simplify the reaction to the smallest whole number coefficients.

If the reaction must occur in basic solution, use step 8.

8. Add the same number of hydroxide ions to both sides of the reaction as there are hydrogen ions. Combine all hydrogen ions with hydroxide ions to make water molecules. Cancel any like items and simplify if possible.

Key 81 Galvanic electrochemical cells

OVERVIEW *An electrochemical cell is a device by which chemical energy can be converted to electrical energy (a galvanic cell), or electrical energy can be converted to chemical energy (an electrolytic cell).*

An **electrochemical cell** is a device in which a redox reaction occurs such that electrons are constrained to move through an external circuit.

A **galvanic** (or **voltaic**) cell is an electrochemical cell that converts chemical energy to electrical energy.

A **half-cell** is one half of an electrochemical cell that is physically separated from the other half by a device that allows completion of the electric circuit but prevents mixing of each half-cell solution.

An **electrolyte** is a solution or a solid that is an ionic conductor.

A **salt bridge** is a tube containing an electrolyte that connects (but separates) the two half-cells of an electrochemical cell.

Electrodes are conductors in an electrochemical cell that make contact with the reactive species.

The **anode** is the positive electrode at which oxidation takes place.

The **cathode** is the negative electrode at which reduction takes place.

Each half-cell has a **half-cell** or **electrode potential**, the potential (or voltage) of one electrode relative to the other.

The **oxidation half-cell potential** ($\mathcal{E}_{ox}$) is the voltage at the anode (oxidation half-cell electrode).

The **reduction half-cell potential** ($\mathcal{E}_{red}$) is the voltage at the cathode (reduction half-cell electrode).

The **overall cell potential** ($\mathcal{E}$) is the sum of the oxidation half-cell potential and the reduction half-cell potential:

$$\mathcal{E} = \mathcal{E}_{ox} + \mathcal{E}_{red}$$

Key 82 Standard half-cell potentials

OVERVIEW *The voltage of an electrochemical cell is dependent on the concentrations of the reactants and products. As a result, standard conditions must be defined. This allows the tabulation of standard half-cell potentials that can be used to predict the direction of an electrochemical reaction.*

The **standard potential** ($\mathcal{E}°$) is the potential of a galvanic cell under standard conditions, that is, T is equal to 25°C, all concentrations are equal to 1.00 M, and the partial pressure of any gas is equal to 1.00 atm.

The **standard hydrogen electrode** is a half-cell used as a reference electrode:

$$2\,H^+(aq) + 2\,e^- \rightarrow H_2(g)$$

Under standard conditions ([H^+] = 1.00 M and P_{H_2} = 1.00 atm), the standard potential for this half-cell is defined as exactly 0.000 V.

A **standard oxidation potential** ($\mathcal{E}°_{ox}$) is the potential of a given anodic half-reaction relative to the hydrogen electrode under standard conditions.

A **standard reduction potential** ($\mathcal{E}°_{red}$) is the potential of a given cathodic half-reaction relative to the hydrogen electrode under standard conditions.

Standard potentials are tabulated as reduction potentials. The following is an abbreviated **table of standard reduction potentials**:

					$\mathcal{E}°_{red}$
$Li^+(aq)$	+	e^-	$\rightarrow$	$Li(s)$	−3.048
$Mg^{2+}(aq)$	+	$2e^-$	$\rightarrow$	$Mg(s)$	−2.357
$Zn^{2+}(aq)$	+	$2e^-$	$\rightarrow$	$Zn(s)$	−0.762
$Co^{2+}(aq)$	+	$2e^-$	$\rightarrow$	$Co(s)$	−0.282
$H_2(g)$	+	$2e^-$	$\rightarrow$	$2\,H^+(aq)$	0.000
$Cu^{2+}(aq)$	+	$2e^-$	$\rightarrow$	$Cu(s)$	0.339
$Ag^+(aq)$	+	e^-	$\rightarrow$	$Ag(s)$	0.799
$Cl_2(g)$	+	$2e^-$	$\rightarrow$	$2\,Cl^-(aq)$	1.360
$F_2(g)$	+	$2e^-$	$\rightarrow$	$2\,F^-(aq)$	2.889

The **overall potential of a cell** ($\mathcal{E}°$) is the sum of the oxidation half-cell potential and the reduction half-cell potential:

$$\mathcal{E}° = \mathcal{E}°_{ox} + \mathcal{E}°_{red}$$

Key 83 Spontaneity of oxidation-
reduction reactions

OVERVIEW *The sign of the overall cell voltage, which is related to the free energy change for the reaction, provides an indication of whether or not an oxidation-reduction reaction will proceed spontaneously.*

An electrochemical reaction will **proceed spontaneously** if the overall cell voltage is positive. If the calculated voltage is negative, the reverse reaction will occur because reversing the reaction changes the sign of the cell potential. Under standard conditions, this means that if

- $\mathcal{E}° > 0$, the reaction is spontaneous.
- $\mathcal{E}° < 0$, the reaction is nonspontaneous.
- $\mathcal{E}° = 0$, the reaction is at equilibrium.

The change in free energy is also the criterion for spontaneity. An equation relating the overall standard cell voltage $\mathcal{E}°$ and the standard free energy change $\Delta G°$ for a voltaic redox reaction can be written as follows:

$$\Delta G° = -n\mathcal{F}\mathcal{E}°$$

n is the number of moles of electrons transferred in the reaction, and $\mathcal{F}$ is the Faraday constant (96,485 C/mol; Key 85).

Inserting the necessary conversion factors, we get the following relation between cell voltage (in V) and $\Delta G°$ (in kJ):

$$\Delta G° \, (\text{kJ}) = -96.5n\mathcal{E}°$$

Note that $\Delta G°$ and $\mathcal{E}°$ have opposite signs; a spontaneous reaction has negative free energy change and positive voltage.

In Key 73, we wrote a relationship between $\Delta G°$ and K, the equilibrium constant:

$$\Delta G° = -RT \ln K$$

Combining this with the above equation gives

$$n\mathcal{F}\mathcal{E}° = +RT \ln K$$

Or, by rearranging, converting the natural logarithm (ln) to a base-10 logarithm (log), and substituting $R = 8.314$ J/mol-K and $\mathcal{F} =$

96,500 C/mol, at $T = 298$ K, we get an **equation relating standard cell voltage and equilibrium constant**:

$$\mathcal{E}° = \frac{0.0591}{n} \log K$$

KEY PROBLEM

The table of standard half-cell potentials (Key 82) can be used to predict whether an oxidation-reduction reaction will proceed spontaneously. As an example, consider this reaction:

$$2\,Ag^+(aq) + Cu(s) \rightarrow 2\,Ag(s) + Cu^{2+}(s)$$

The relevant reduction potentials are

$$Cu^{2+}(aq) + 2\,e^- \rightarrow Cu(s) \qquad \mathcal{E}° = 0.339 \text{ V}$$
$$Ag^+(aq) + e^- \rightarrow Ag(s) \qquad \mathcal{E}° = 0.779 \text{ V}$$

Since the overall potential of the cell is given by the sum of the oxidation half-cell potential and the reduction half-cell potential, we must reverse the reduction half-cell for Ag, which is being oxidized. Note that when the cell is reversed, the sign of the potential is reversed.

$$Cu(s) \rightarrow Cu^{2+}(aq) + 2\,e^- \qquad \mathcal{E}° = -0.339 \text{ V}$$
$$Ag^+(aq) + e^- \rightarrow Ag(s) \qquad \mathcal{E}° = 0.779 \text{ V}$$

Adding these two half-cell reactions (and potentials) gives the equation for the complete reaction, as well as the overall cell potential (0.440 V). Note that since we are dealing with electrochemical potential (a potential energy), even though the Ag half-cell had to be multiplied by 2 to balance the overall equation, we do not multiply the half-cell reaction by the number of electrons transferred. Furthermore, since the cell potential is positive, the reaction will proceed spontaneously.

We can then use the equation relating $\mathcal{E}°$ to K to determine the equilibrium constant for this reaction under standard conditions. Rearranging that equation and using $n = 2$ (two electrons are transferred in the electrochemical reaction), we get

$$\log K = \frac{n\mathcal{E}°}{0.0591}$$
$$\log K = \frac{(2)(0.440)}{0.0591} = +14.89$$
$$K = 7.76 \times 10^{14}$$

Key 84 The Nernst equation

OVERVIEW *Because voltaic cells are unlikely to operate only under standard conditions, it is necessary to consider what a cell voltage would be under nonstandard conditions. The relationship between cell voltage and concentration is known as the Nernst equation.*

In Key 73, we wrote an expression relating ΔG°, ΔG, and Q, the numerical value of the mass action expression:

$$\Delta G = \Delta G^\circ + RT \ln Q$$

or, in terms of a base-10 logarithm,

$$\Delta G = \Delta G^\circ + 2.30 \, RT \log Q$$

Substituting $-n\mathcal{F}\mathcal{E}$ for ΔG into this equation and rearranging gives

$$\mathcal{E} = \mathcal{E}^\circ - \frac{2.30^\circ \, T}{nF} \log Q$$

This is known as the **Nernst equation** and relates standard half-cell potentials and concentrations to half-cell voltage.

Substituting for the constants at 25°C (298 K), the **Nernst equation** reduces to

$$\mathcal{E} = \mathcal{E}^\circ - \frac{0.0591}{n} \log Q$$

Note that when the system is at equilibrium, $Q = K$ and $\mathcal{E} = 0$, the Nernst equation reduces to

$$\mathcal{E}^\circ = \frac{0.0591}{n} \log K \qquad \text{(Key 83)}$$

Key 85   Electrolytic cells

OVERVIEW *The reverse of a voltaic cell is an electro-chemical cell that converts electrical energy to chemical energy. One common use of an electrochemical cell is in the deposition of material from solution using an electric current.*

An **electrolytic cell** is an electrochemical cell that converts electrical energy to chemical energy.

Electrolysis involves the passage of an electric current through a solution in order to force a chemical reaction to proceed.

Faraday's law relates the amount of material produced and consumed to the amount of charge passing through an electrochemical cell. The **Faraday** ($\mathscr{F}$) contains one mole of e^- and equals 96,500 coulombs (C). 1 $\mathscr{F} = 1\,\text{mol}\,e^- = 96{,}500$ C. In electrochemistry, one **equivalent** is the amount of substance that gains or produces one mole of e^-.

the **ampere** is the unit of electric current having the units C/s, or $1\,\text{C} = 1\,\text{amp-s}$.

Using this information, one can calculate the amount of a metal, for example, deposited from solution by electrolysis. If 10.0 amp of current passed through an electrolytic cell containing Vanadium (III), V^{3+}, for 5.00 min, how many grams of Vanadium metal will be deposited?

First convert 10.0 amp flowing for 5.00 min to coulombs:

$$5.00\,\text{min} \times \frac{60\,\text{s}}{1\,\text{min}} \times 10.0\,\text{amp} = 3.00 \times 10^3\,\text{C}$$

Next, convert coulombs to Faradays:

$$(3.00 \times 10^3\,\text{C}) \times \frac{1\,\mathscr{F}}{96{,}500\,\text{C}} = 3.11 \times 10^{-2}\,\mathscr{F}$$

The redox equation ($V^{3+} + 3e^- \rightarrow V$) states that $3\,\text{mol}\,e^-$ is equivalent to $1\,\text{mol}\,V$. Therefore,

$$(3.11 \times 10^{-2}\,\text{mol}\,e^-) \times \frac{1\,\text{mol}\,V}{3\,\text{mol}\,e^-} \times \frac{50.94\,\text{g}\,V}{1\,\text{mol}\,V} = 0.528\,\text{g}\,V$$

A **primary cell** is an electrochemical cell that can act only as a galvanic cell (it cannot be recharged).

A **secondary cell** is an electrochemical cell that can act as a galvanic cell and as an electrolytic cell (it can be recharged).

Theme 18 ORGANIC CHEMISTRY

*O*rganic chemistry deals with the chemistry of the element carbon, which forms an extremely large number (perhaps as many as 90%) of known compounds. This theme will provide a brief outline of carbon chemistry.

Key 86   Alkanes

OVERVIEW *The extensive chemistry of carbon results from the ability of each carbon atom to form up to four bonds with other atoms as well as to link together to form chains or rings. This Key introduces alkanes, which contain carbon-carbon single bonds.*

A **hydrocarbon** is a compound that contains only two kinds of atoms, hydrogen and carbon.

An **aliphatic compound** is composed of chains of carbon atoms or nonaromatic (Key 88) rings of carbon atoms.

A **cyclic aliphatic compound** is a compound that has at least one nonaromatic ring.

In aliphatic hydrocarbons, each **carbon atom is bonded tetrahedrally** to four other atoms, forming a tetrahedron with a carbon atom at the center.

This tetrahedral bonding is the result of sp^3 **hybridization** (Key 36) about each carbon atom, allowing the formation of four single bonds.

Alkanes, also known as **olefins** or **saturated hydrocarbons**, are hydrocarbons that have no double bonds or triple bonds; that is, the maximum number of hydrogen atoms is attached to the carbon atoms.

An **alkyl group** is a hydrocarbon minus one hydrogen atom. The ethane molecule CH_3CH_3 becomes the ethyl group $-CH_2CH_3$.

The nomenclature for **straight-chain alkanes** is as follows: methane (CH_4); ethane (CH_3CH_3); propane ($CH_3CH_2CH_3$); butane [$CH_3(CH_2)_2CH_3$]; pentane [$CH_3(CH_2)_3CH_3$]; hexane [$CH_3(CH_2)_4CH_3$]; and so on.

For **branched-chain alkanes**, each name is derived by naming the longest straight-chain portion, replacing one or more hydrogens by alkyl groups, and adding a prefix indicating the alkyl group. For example,

$$CH_3-\underset{\overset{|}{CH_3}}{CH}-CH_2-CH_3$$

is named 2-methylbutane, the 2 indicating that the methyl group is attached to the second carbon from the end.

Key 87   Alkenes and alkynes

OVERVIEW *Alkenes and alkynes are members of the class of compounds called unsaturated hydrocarbons, compounds in which one or more of the carbon-carbon bonds is a multiple bond.*

An **unsaturated hydrocarbon** is a hydrocarbon that has double or triple bonds; that is, additional hydrogen atoms can be added to the carbon atoms.

An **alkene** is a hydrocarbon that contains at least one carbon-carbon double bond (C═C) but no triple bonds.

The carbon atoms involved in the double bond are sp^2 **hybridized** (Key 36), with three planar triangular sp^2 hybrids around the carbon. One of these hybrids, along with the remaining p orbital, forms the double bond. The other two hybrid orbitals form C—X bonds with a bond angle of 120° and restricted rotation about the double bond.

An **alkyne** is a hydrocarbon that contains at least one carbon-carbon triple bond (C≡C).

The carbon atoms involved in the triple bond are sp **hybridized** (Key 36), with two linear sp hybrids about the carbon. One of these hybrids, along with the remaining two p orbitals, forms the triple bond. The remaining hybrid orbital forms a C—X bond leading to linear coordination with no rotation about the triple bond.

The **nomenclature** for the simplest hydrocarbons uses a suffix to differentiate among alkanes (suffix **-ane**), alkenes (suffix **-ene**), and alkynes (suffix **-yne**). For example, for eth**ane** the formula is CH_3—CH_3, eth**ene** (also known as ethylene) is CH_2═CH_2, and eth**yne** (also known as acetylene) is CH≡CH.

Alkenes and alkynes can be converted to the corresponding alkane by **hydrogenation**, the catalytic addition of hydrogen across all the multiple bonds:

$$CH_2═CH—CH_2—C≡CH + 3\ H_2 \rightarrow CH_3—(CH_2)_3—CH_3$$

Key 88 Aromatic hydrocarbons

OVERVIEW *All aromatic compounds contain the benzene ring [or a system containing (4n + 2) π electrons], a structure that can be described as having more than one resonance structure.*

An **aromatic compound** is a compound that contains a ring that can be represented by more than one resonance structure.

Benzene (C_6H_6) the prototypic aromatic compound, is a planar molecule. The first rationalization of its structure was a resonance hybrid of two equivalent Lewis structures with alternating single and double carbon-carbon bonds (**Kekule structures**):

In Key 38, we introduced **delocalized molecular orbitals**, in which the electrons in a bond are spread out over several atoms. To explain the bonding in benzene in these terms, first consider each carbon to be sp^2 hybridized. Six C—C σ bonds and six C—H σ bonds are formed from these hybid orbitals, leaving six $2p$ orbitals (one on each carbon) oriented perpendicularly to the plane of the ring. These orbitals are occupied by six electrons forming three π bonds that are delocalized over the entire ring and are usually schematically indicated as follows:

There are a large number of **derivatives of benzene** that consist of one or more functional groups (Key 89) substituted for a hydrogen atom on the ring. Chlorobenzene (C_6H_5Cl) and nitrobenzene ($C_6H_5NO_2$) are examples of monosubstituted benzenes. With two groups on the ring, three isomers (Key 90) are possible. The isomers of dichlorobenzene ($C_6H_4Cl_2$) are named with the prefixes *ortho-* (*o-*), *meta-* (*m-*), and *para-* (*p-*):

| *o*-dichlorobenzene | *m*-dichlorobenzene | *p*-dichlorobenzene |
| (1,2-dichlorobenzene) | (1,3-dichlorobenzene) | (1,4-dichlorobenzene) |

Another class of aromatic compounds is based on **condensed benzene rings** and consists of two or more benzene rings fused together:

naphthalene ($C_{10}H_8$)

anthracene ($C_{14}H_{10}$)

Key 89 Functional groups

OVERVIEW *The bulk of the chemistry of carbon compounds revolves around the reactivity of functional groups. Furthermore, carbon compounds can be classified into families of compounds according to the nature of their functional groups.*

A **functional group** is either a nonmetal atom or group of atoms bonded to carbon (and replacing a hydrogen atom).

An **alkyl halide** is a compound composed of an alkyl group and a halide, for example, methyl chloride, $-CH_3Cl$. In alkyl halides, the halogen atom $-X$ is considered to be the functional group.

A **primary alkyl halide** is a compound in which the carbon atom to which the halogen is attached is joined to only one other carbon atom.

A **secondary alkyl halide** is a compound in which the carbon atom to which the halogen is attached is joined to two other carbon atoms.

A **tertiary alkyl halide** is a compound in which the carbon atom to which the halogen is attached is joined to three other carbon atoms.

An **alcohol** is an organic compound of the family R—OH. In this case, the hydroxyl group, —OH, is the functional group.

An **aldehyde** is an organic compound of the family $R-\overset{\displaystyle O}{\overset{\displaystyle \|}{C}}-H$.

A **ketone** is an organic compound of the family $R\overset{\displaystyle O}{\overset{\displaystyle \|}{C}}R$.

The **carboxyl group** is the functional group $-\overset{\displaystyle O}{\overset{\displaystyle \|}{C}}-OH$.

A **carboxylic acid** is a compound of the family $R\overset{\displaystyle O}{\overset{\displaystyle \|}{C}}-OH$.

An **ester** is an organic compound of the family $R\overset{\displaystyle O}{\overset{\displaystyle \|}{C}}-OR$.

An **ether** is an organic compound of the family ROR.

The **amide group** is the functional group $-\overset{\displaystyle O}{\overset{\displaystyle \|}{C}}-NH_2$.

The **amine group** is the functional group $-NH_2$.

Key 90 Isomers and optical activity

OVERVIEW *Isomers are different compounds (having different properties) with the same molecular formula. One class of isomers differs because the tetrahedral bonding around the carbon atom can result in mirror images of a molecule.*

Isomers are compounds that have the same molecular formula but different molecular structures.

Structural isomers are isomers differing in the sequence in which the atoms are bonded together. For example, two structural isomers with the empirical formula C_2H_6O are methyl ether (CH_3—O—CH_3) and ethyl alcohol (CH_3CH_2OH).

Geometric isomers are isomers that occur when two species with the same molecular formula have different geometric structures. The three isomers of dichloroethene are

1,1-dichloroethene   *cis*-1,2-dichloroethene   *trans*-1,2-dichloroethene

The cis isomer has the substitute groups on the same side of the double bond (as close to each other as possible); the trans isomer has the substitute groups on opposite sides of the double bond.

Optical isomers are isomers that are nonsuperimposable mirror images of each other. These are known as **enantiomers**.

An **asymmetric carbon atom** is one that is bonded to four different groups and is a center of chirality.

A **chiral molecule** possesses a "handedness," as applied to the case of optical isomers, which are not superimposable on their mirror images.

Optical activity is the rotation of the plane of polarized light as it passes through a chiral substance.

A **levorotatory isomer** is an optical isomer that causes rotation of the plane of polarized light in a clockwise direction.

A **dextrorotatory isomer** is an optical isomer that causes rotation of the plane of polarized light in a counterclockwise direction.

A **racemic mixture** is a mixture containing equal numbers of the two optical isomers of a substance.

GLOSSARY

Included here are the definitions of many, but not all, of the terms used in the Keys. For terms not listed here, please consult the index.

absolute zero The lowest temperature attainable; $-273.15°C$; zero on the Kelvin temperature scale.

accuracy of a measurement Tells you how close the measurement is to the true (or absolute) value of the quantity.

acid A substance that increases the concentration of H^+ (or H_3O^+) in solution. Once was defined as a compound that tastes sour, feels prickly on the skin, turns the organic dye litmus red, and reacts with certain metals to produce hydrogen gas.

acid anhydride A nonmetal oxide that reacts with water to form an acid.

actinides (or **transuranium elements**) The elements located after uranium in the periodic table (actually thorium through lawrencium); formed by filling the $5f$ sublevel.

activated complex (or **transition state**) An intermediate species found during a chemical reaction when reactants have collided and are at the high point on the potential energy diagram for the reaction.

activation energy (E_a) The minimum energy that must be surpassed in order for a chemical reaction to take place; it also can be considered as the potential energy barrier to the reaction.

alcohol An organic compound of the family R—OH.

aldehyde An organic compound of the family $R-\overset{\overset{O}{\|}}{C}-H$.

aliphatic compound A compound composed of chains of carbon atoms; cyclic aliphatic compounds have at least one nonaromatic ring.

alkali metals Elements in Group IA (or 1) of the periodic table.

alkaline earth metals Elements in Group IIA (or 2) of the periodic table.

alkane (olefin, saturated hydrocarbon) A hydrocarbon that has no double or triple bonds; the maximum number of hydrogen atoms are attached to the carbon atoms.

alkene A hydrocarbon that contains at least one carbon-carbon double bond ($C=C$) but no triple bonds.

alkyl group A hydrocarbon minus one carbon atom, such as the methyl group $-CH_3$.

alkyne A hydrocarbon that contains at least one carbon-carbon triple bond ($C\equiv C$).

amide group The functional group $-\overset{\overset{O}{\|}}{C}-NH_2$.

amine group The functional group $-NH_2$.

amphoteric substance A compound that can behave as either an acid or a base.

ångstrom (Å) $1 Å = 1.00 \times 10^{-8}$ cm.

anion A negatively charged ion.

anode The positive electrode at which oxidation takes place.

aromatic compound A compound that contains a ring that can be represented by more than one resonance structure.

Arrhenius acid and base An Arrhenius acid is a compound that dissociates in aqueous solution to give H^+ (or H_3O^+) ions; an Arrhenius base is a compound that dissociates to give OH^- ions.

Arrhenius equation An equation relating the rate constant for a reaction k to absolute temperature T: $k = Ae^{-(E_a/RT)}$, where A is a constant related to the frequency of each collision, E_a is the activation energy, and R is the gas constant (8.314 J/mol-K).

atmospheric pressure The pressure exerted by the atmosphere; equals the pressure exerted by a column of mercury (Hg) 760 mm high; 1 atm = 760 mm Hg or 760 torr.

atom The smallest unit of an element having all the characteristics of that element.

atomic mass number (A) A number equal to the number of protons plus the number of neutrons in the nucleus of an atom.

atomic mass unit (amu) A mass equal to $1/12$ the mass of an atom of ^{12}C; 1.66053×10^{-24} g.

atomic number (Z) A number equal to the number of protons in the nucleus of an atom; Z uniquely determines the identity of an element.

atomic radius The effective radius of an atom measured in its elemental state (one-half the interatomic distance). Atomic radii tend to increase going down a column in the periodic table and to decrease going across a row (from left to right).

atomic weight (or atomic mass) The mass of an atom of an element expressed in atomic mass units.

aufbau principle Used to build up electronic configurations of the elements; as protons are individually added to the nucleus to build up the elements, electrons are similarly added to the atomic orbitals.

autoionization The process in which water self-dissociates: $H_2O + H_2O \rightleftarrows H_3O^+(aq) + OH^-(aq)$

Avogadro's law At constant temperature and pressure, the volume occupied by a gas is proportional to the number of moles of gas present; $V_{(T, P)} = kn$, where k is a proportionality constant. This is also stated as Avogadro's principle: Equal volumes of different gases at the same temperature and pressure contain the same number of molecules.

Avogadro's number The number of atoms, molecules, or formula units per mole (or gram-atom or gram-formula weight, respectively) of a substance; numerically equal to 6.022×10^{23}.

azimuthal (or angular momentum) quantum number (l) The quantum number that defines the type of electronic orbital; can have values of 0, 1, 2, 3, . . . $(n - 1)$. Values of these quantum numbers correspond to s, p, d, and f orbitals for $l = 0$, 1, 2, and 3, respectively.

barometer An instrument used to measure atmospheric pressure.

base A substance that increases the concentration of OH^- in solution. Historically defined as a compound that tastes bitter, feels slippery, and turns litmus blue.

base anhydride A metal oxide that reacts with water to form a base.

berthollides Nonstoichiometric solids in which the mole ratios of the component elements are not small whole numbers.

Bohr quantum number The principal quantum number n; can have only integral values.

boiling point The temperature at which a liquid changes to a gas.

boiling point elevation The increase in the boiling point of a solution above the boiling point of the pure solvent; equal to the boiling point elevation constant (K_b, characteristic of the solvent) times the molality of the solution.

bond dipole A dipole formed between two atoms involved in a bond caused by unequal sharing of the electrons in the bond, thus making one end of the bond positive and the other end negative.

bond energy The amount of energy needed to separate two atoms joined by a chemical bond.

bond length The distance between the nuclei of two atoms joined by a chemical bond.

bond order The net number of pairs of bonding electrons, equal to the number of electrons in bonding MOs minus the number of electrons in antibonding MOs, divided by 2.

Boyle's law The volume of a given amount of gas held at a constant temperature is inversely proportional to pressure; $V_{(n,T)} = k(1/P)$

Brønsted-Lowry acid and base A Brønsted-Lowry acid is a species that donates a proton to another species in a reaction; a Brønsted-Lowry base is a species that accepts a proton. A conjugate acid-base pair consists of a Brønsted acid (HX) and the base (X^-) formed by the removal of a proton from that acid; X^- is the conjugate base of HX.

Brownian motion The continuous random motion observed in particles suspended in a gas or a fluid.

buffer solution A solution made from a weak acid and a salt whose anion is the conjugate base of that acid (or a weak base and the conjugate acid of that base); a solution resistant to large changes in pH.

buret A graduated glass tube fitted with a valve at one end that is used to dispense measured volumes of liquid.

calorie (cal) A unit of thermal energy; the amount of heat required to raise the temperature of 1 g of water by 1°C (more exactly, from 14.5°C to 15.5°C).

calorimeter A device used to measure energy changes associated with a

physical or a chemical process. In an ordinary calorimeter (at constant pressure), the heat flow measured, $q_{reaction} = \Delta H$ for the reaction. In a bomb calorimeter (which is held at constant volume), the heat measured, $q_{reaction} = \Delta E$.

carboxylic acid A compound of the family $R-\overset{\overset{\displaystyle O}{\|}}{C}-OH$. Contains the carboxyl functional group $-\overset{\overset{\displaystyle O}{\|}}{C}-O-$.

catalyst A substance that speeds up or slows down a chemical reaction without being consumed. A catalyst acts by lowering the activation energy for a reaction.

cathode The negative electrode at which reduction takes place.

cathode rays Emitted by the cathode (negative electrode) in a gas discharge tube. Investigated by J. J. Thomson, who determined their charge-to-mass ratio. They consist of a stream of electrons.

cation A positively charged ion.

Celsius scale (°C) The temperature scale on which 0°C is defined by the freezing point of water and 100°C is defined by the boiling point of water.

chain reaction A reaction in which the product of one step is the reactant in another step.

chalcogens (or **chalcogenides**) Elements in Group VIA (or 16) of the periodic table.

Charles' law The volume of a given amount of gas held at constant pressure is directly proportional to the absolute temperature; $V_{(n,P)} = kT$.

chemical equation A chemical shorthand notation that describes a reaction between chemical substances and provides the necessary information for calculating weight relations in the reaction. The coefficients in a balanced chemical equation represent the relative numbers of moles of reactants and products entering into that reaction.

chemical equilibrium A dynamic equilibrium that involves a chemical reaction.

chemical formula A representation of the elemental composition of a compound, in which the subscripts following each element indicate the atomic ratio of each element in the compound.

chemical property A property that can be studied only at the risk of changing the identity of the substance.

chiral molecule A molecule possessing a handedness, as in the case of optical isomers, which are not superimposable on their mirror images.

cleavage In most crystalline solids, the separation along specific planes when subjected to external mechanical stress to give fragments with flat surfaces dependent on the geometric array and type of bonding.

colligative properties Properties of solutions that result from the presence of a solute but are independent of that specific solute; they include vapor pressure lowering, boiling point elevation, freezing point depression, and osmotic pressure.

collision theory A theory of reaction rates that states that the rate of a reaction is proportional to the number of collisions per second between reactant molecules.

colloid (or **colloidal suspension**) A mixture in which the particles of the lesser phase are intermediate in size between those present in homogeneous mixtures and those present in heterogeneous mixtures. These particles are typically on the order of 10 to 10,000 Å in diameter.

common ion effect A shift in equilibrium caused by the addition of a compound having an ion that is common to the dissolved species.

compound A chemical combination of elements; a pure substance, uniquely characterized by its properties, which are different from those of the component elements, and in which the component elements are present in fixed proportions.

condensation point The temperature at which a gas changes to a liquid (the same temperature as the boiling point).

covalent compound A compound that contains only covalent bonds between two atoms. Covalent bonds are formed by sharing one or more pairs of electrons.

covalent network solids Solids in which atoms form strong covalent bonds in an infinite three-dimensional network of covalently bonded species. Generally, they have high melting and boiling points and are formed between elements near the Zintl border in the center of the periodic table.

critical pressure (P_c) The pressure that liquefies a gas at its critical temperature.

critical temperature (T_c) The temperature above which thermal motion is so violent that a gas cannot be liquefied; the liquid state of a substance cannot exist above T_c for that substance.

crystal lattice The repeating pattern of atoms, molecules, or ions in a crystal, that is, the three-dimensional array of lattice points.

crystalline solid A solid that has an ordered internal arrangement of atoms, ions, or molecules that vibrate about fixed positions in an ordered geometric array.

Dalton's atomic theory An early theory of atomic structure that states that the basic unit of an element is an extremely small, indivisible particle called an atom, a hard sphere with a characteristic size and mass. Each atom of the same element has the same mass; atoms of one element are different from atoms of another element. Atoms of one element can combine with atoms of a second element, usually in small whole number ratios, to form molecules—the characteristic unit of a compound.

Dalton's law of partial pressures In a closed container, the total pressure is the sum of the partial pressures of the individual gases; $P_T = \Sigma P_i$, where P_T is the total pressure, P_i represents the individual partial pressures, and Σ denotes "sum of."

daltonides Stoichiometric solids in which the mole ratios of the component elements of a substance are small whole numbers.

degenerate orbitals Electronic orbitals that have the same energy.

density The ratio of the mass of an object to the volume occupied by that object; $D = m/V$, usually expressed in g/cm^3.

dextrorotatory isomer An optical isomer that causes rotation of the plane of polarized light in a counter-clock-wise direction.

dilution The process of preparing a less concentrated solution from a more concentrated one.

dimensional (or **factor-label**) **analysis** The use of appropriate conversion factors to solve problems.

dipole moment The product of the partial charge on either end of a dipole multiplied by the distance between the charges.

dipole-dipole interactions Intermolecular interactions between polar molecules.

disproportionation The process in which the same chemical substance undergoes both oxidation and reduction.

dissociation Ionic compounds dissolve with the formation of ions in solution.

Dobereiner triads A numerical relationship among atomic weights of certain groups of three elements having similar properties.

double bond A covalent bond in which two pairs of electrons are shared.

dynamic equilibrium A system in which the rates of the forward and reverse reactions are equal.

electric dipole Results from a separation of positive and negative charges and has both magnitude and direction.

electrochemical cell A device in which a redox reaction occurs such that electrons are constrained to move through an external circuit.

electrode A conductor in an electrochemical cell that makes contact with the reactive species.

electrolysis The process that involves the passage of an electric current through a solution in order to force a chemical reaction to proceed.

electrolyte A solution or a solid that is an ionic conductor.

electrolytic cell An electrochemical cell that converts electrical energy to chemical energy.

electromagnetic radiation Can be considered to be composed of waves that travel at the speed of light ($c = 3.00 \times 10^{10}$ cm/s). These waves are characterized by a wavelength (λ) and a frequency (v); the relationship between wavelength and frequency is given by λ (cm) $\times$ v (s^{-1}) $= c$ (cm/s). The electromagnetic spectrum consists of all electromagnetic radiation, from the low ν (long λ) radio waves through the visible range and into the high ν (short λ) cosmic ray region.

electron A charged particle (symbol e) that carries one negative charge (-1.602×10^{-19} coulomb, or 1 electronic charge unit, esu) but weighs only about 1/2000 of an amu.

electron affinity The energy emitted in the process of adding an electron to an atom (resulting in the formation of a negative ion).

electron-deficient molecules Molecules that do not conform to the octet rule (i.e., those having less than eight valence electrons).

electronegativity (EN) The tendency of an atom to attract a shared pair of electrons in a chemical bond. The electronegativity scale is normalized to elemental fluorine which is assigned EN = 4.0. The general trend in electronegativity increases going from left to right in the periodic table and from bottom to top in a given group. ΔEN (the difference in electronegativity) between two atoms in a bond is a measure of the amount of ionic character in that bond.

electron spin A property of an electron resulting from the fact that an electron can appear to act as a tiny spinning sphere of negative charge generating a magnetic field.

electronic configuration The distribution of electrons in the various energy levels in an atom.

electronic orbital The region in space around a nucleus in which there is maximum probability of finding the electron, that is, a plot of Ψ^2.

element A pure substance that cannot be chemically separated or decomposed into simpler substances. There are 103 known elements.

elementary process One of the individual steps in a multistep reaction mechanism.

empirical formula The simplest chemical formula that expresses the relative number of moles of elements in a compound using the smallest whole numbers.

endothermic process A process that takes place with the absorption of heat; ΔH is positive.

endpoint (or **equivalence point**) The point in a titration when equal numbers of equivalents of reactants have been combined, that is, the completion of the reaction.

enantiomers See optical isomers.

energy level diagram A diagram that illustrates the electronic orbitals found at different energies.

enthalpy (***H***) A measure of the heat content of a substance.

entropy (*S*) A measure of the disorder of a system. The standard entropy ($S°$) of a substance is the entropy of one mole of that substance at 25°C and 1 atm. For a reaction or process, the standard entropy change $\Delta S° = (\Sigma \Delta S° \text{ products}) - (\Sigma \Delta S° \text{ reactants})$.

equation of state An equation that describes the physical behavior of a state of matter.

equilibrium A condition that exists when there are no observable changes in the system with time.

equilibrium constant (***K***) An experimentally determined number that describes the relative concentrations of reactants and products in a particular equilibrium; it is dependent on temperature. The magnitude of the equilibrium constant is a measure of the position of the equilibrium.

equilibrium constant for an acid (***K***$_a$) The equilibrium expression for the dissociation of an acid in aqueous solution.

equilibrium constant for a base (K_b) The equilibrium expression for the dissociation of a base in aqueous solution.

ester An organic compound of the family RCOR.
$$\begin{array}{c} O \\ || \end{array}$$
family RCOR.

ether An organic compound of the family ROR.

evaporation The process by which the molecules that have the most kinetic energy escape from the surface of a liquid; an example of a dynamic equilibrium between a liquid and a gas.

excited state An electron in an orbit (state) or higher energy.

exothermic process A process that takes place with the evolution of heat; ΔH is negative.

expanded octet The electronic configuration found in a molecule in which the central atom is in groups V through VIII and in the third, fourth, and fifth periods where up to six pairs of electrons (12 valence electrons) can be held.

extensive property A property that depends on the amount of material present, for example, mass and volume.

Fahrenheit scale (°F) The temperature scale on which 32°F is defined by the freezing point of water and 212°C is defined by the boiling point of water.

Faraday ($\mathcal{F}$) One mole of e^-; 1 $\mathcal{F}$ = 96,500 Coulombs.

first law of thermodynamics When a system undergoes a series of changes that eventually bring it back to its original state, the net energy change is zero; $\Delta E = q + w$.

formal charge The difference between the number of valence electrons in an isolated atom and the number of electrons assigned to that atom in a Lewis structure.

formula weight (or molecular weight) The sum of the masses of the atoms in the chemical formula of a compound.

freezing point The temperature at which a liquid changes to a solid (the same temperature at the melting point).

freezing point depression The decrease in the freezing point of a solution below the freezing point of the pure solvent; equal to the freezing point depression constant (K_f, characteristic of the solvent) times the molality of the solution.

frequency (v, nu) The number of times a wave moves past a given reference point within a unit of time; number of waves or cycles per second.

functional group Either a nonmetal atom or group of atoms bonded to carbon (replacing a hydrogen atom).

galvanic (or voltaic) An electrochemical cell that converts chemical energy to electrical energy.

gas constant (R) The proportionality constant appearing in the ideal gas law; has a numerical value of 0.0821 L atom/mol-K (or 8.314 J/mol-K).

gaseous state A totally disordered state of matter in which the molecules move constantly in rapid, random translational motion.

Gay-Lussac's law The pressure exerted by a given amount of gas held at constant volume is directly proportional to the absolute temperature; $P_{(n,V)} = kT$.

geometric isomers Isomers that occur when two species with the same molecular formula have different geometric structures.

Gibbs free energy (G) The thermodynamic function that relates enthalpy (H) and entropy (S); $G = H - TS$, or for a process, $\Delta G = \Delta H - T\Delta S$. Under standard conditions, $\Delta G° = \Delta H° - T\Delta S°$; for a reaction, the standard free energy change $\Delta G° = (\Sigma\Delta G°$ products) $- (\Sigma\Delta G°$ reactants).

glass (or amorphous solid) A permanently supercooled liquid with an internal structure characteristic of a liquid not a crystalline solid.

graduated cylinder A cylinder used to measure given amounts of liquid.

Graham's law of diffusion The rate of diffusion of a gas is inversely proportional to the square root of its molecular weight. When comparing the rates of diffusion of two gases (A and B), the ratio $\dfrac{\text{rate}_A}{\text{rate}_B} = \left(\dfrac{\text{FM}_B}{\text{FM}_A}\right)^{1/2}$, where FM_A and FM_B are the formula weights of A and B, respectively.

gram-atom That amount of an element that has the same mass in grams as the atomic mass in amu. The gram-formula weight is the amount of a compound that has the same mass in grams as the formula mass in amu.

ground state The lowest energy electronic configuration in an atom.

group (or family) of elements A vertical column of elements in the periodic table.

habit The external shape and form of a crystal; characteristic faces define the crystal and characteristic angles between the faces.

half-cell One-half of an electrochemical cell that is physically separated from the other half by a device that allows completion of the electric circuit but prevents mixing of each half-cell solution. The potential (or voltage) of one half-cell electrode relative to the other is called the half-cell potential.

half-life ($t_{1/2}$) The time required for the concentration of a reactant to be reduced to one-half its initial value.

half-reaction Every oxidation-reduction reaction can be separated into two half-reactions, one involving the oxidation process, the other the reduction process.

halogens (or halides) Elements in Group VIIA (or 17) of the periodic table.

heat capacity The amount of heat necessary to raise the temperature of a substance by 1°C.

heat of combustion The amount of heat evolved in the reaction of a compound with oxygen.

heat of fusion (ΔH_f) The amount of heat necessary to convert one mole of a solid to one mole of liquid.

heat of reaction (ΔH) The enthalpy change associated with a reaction.

heat of vaporization (ΔH_v) The amount of heat necessary to convert one mole of a liquid to one mole of gas.

heating curve A plot of temperature versus time when adding heat to a closed system at a constant rate.

Heisenberg's uncertainty principle A fundamental limitation on the measurement of certain pairs of observables, such as position and momentum. As a result, it is impossible to determine both the position and energy of an electron in an atom.

Henry's law The solubility of a gas in a liquid is directly proportional to the partial pressure of the gas over the resultant solution.

Hess's law When thermochemical equations are added, the heats of reaction are added to give the net heat of reaction.

heterogeneous Containing two or more physically discernible phases.

heterogeneous equilibrium An equilibrium that involves two or more phases.

heterogeneous mixture A mixture made up of two or more physically discernible parts or phases.

homogeneous Consisting of similar parts; alike.

homogeneous mixture A mixture in which the components cannot be individually detected.

Hund's rule Given the choice, there is a maximum number of unpaired electrons with parallel spins in an electronic configuration.

hydrocarbon A compound that contains only hydrogen and carbon.

hydrogen bonds Intermolecular forces that result from a specific interaction between a hydrogen atom in one molecule and a fluorine, oxygen, or nitrogen atom in another molecule.

hydrogenation The catalytic addition of hydrogen across all the multiple bonds in alkenes, alkynes, and aromatic compounds.

hydrolysis The reaction of a compound with water, usually resulting in a change in the pH of the solution.

hydronium ion (H_3O^+) The notation for a hydrated proton; that is, a proton in aqueous solution.

hydrophilic A substance that is strongly attracted to water molecules and therefore tends to be soluble in water.

hydrophobic A substance that is very weakly attracted to water molecules and therefore tends to be insoluble in water.

ideal gas A gas that obeys Boyle's law; in terms of the kinetic molecular theory, a gas in which there are no attractive forces between the molecules and in which the molecules are point masses; obeys the ideal gas law; $PV = nRT$.

ideal solution A solution in which there is negligible interaction between solute molecules; ideal solutions obey Raoult's law.

indicator A compound (usually a weak acid) that undergoes a color change with a change in pH; used to determine the endpoint of a titration.

induced dipole A dipole created when the electron cloud is distorted by a neighboring dipole or ion.

instantaneous dipole A dipole created by the motion of electrons in atoms.

instantaneous rate The reaction rate at a particular time; obtained by taking the tangent to a plot of the change in concentration of a reactant or product versus time.

intensive property A property that does not depend on the amount of material present, for example, temperature.

intermolecular forces Forces between the molecules of molecular substances that tend to be weaker than normal covalent bonds (ion-dipole and dipole-dipole interactions, London forces, and hydrogen bonds), as opposed to intramolecular forces, which are forces between the atoms making up molecules.

internal energy (E) The total kinetic and potential energy of a system.

interstitial solid solution A solid solution in which solute particles fit into empty spaces between the particles of the solvent.

ion An atom or group of atoms that has a positive charge or a negative charge.

ion-dipole interactions Intermolecular forces between an ion and a molecule that has a dipole.

ionic bond A bond formed between positively charged ions and negatively charged ions in which the interaction between these species is electrostatic in nature.

ionic radius The size of an ion; for ions of the same charge (i.e., in the same family), the radius increases going down the periodic table.

ionic solids Solids in which the building blocks are positive and negative ions. Ionic solids generally have high melting and boiling points and do not conduct heat or electricity.

ionicity The amount of ionic character in a bond (compared to covalent character).

ionization The process by which a molecular substance separates into ions when dissolved in a polar solvent such as water.

ionization energy (or **ionization potential**) The energy needed to remove an electron from an isolated atom.

isoelectronic species Species (atoms, ions, or molecules) that have the same number of electrons.

isomers Compounds that have the same molecular formula but different molecular structures.

isotonic solutions Solutions that have the same osmotic pressure.

isotopes Atoms of a given element that have the same number of protons in the nucleus (same Z) but a different number of neutrons (different A); they differ in atomic mass.

Kelvin (or **absolute temperature scale**) The temperature scale in which one degree has the same magnitude as one degree on the Celsius scale and the zero point is $0\,K$ ($-273.15°C$).

ketone An organic compound of the family $R{-}\overset{\displaystyle O}{\overset{\displaystyle \|}{C}}{-}R$.

kinetic energy (KE) Energy associated with motion; given by the formula $KE = \frac{1}{2}mv^2$, where m = mass (in g) and v = velocity (or speed, in cm/s). KE has units $g \cdot cm^2/s^2$ or ergs (1 erg = $1\,g \cdot cm^2/s^2$). In the SI, the unit of energy is the joule (J), where 1 joule = 1×10^7 ergs.

kinetic molecular theory A theory used to explain the behavior of gases that assumes that the molecules in a gas are point masses (their volume is negligible compared to the volume of the container) and that there are no attractive forces between the molecules, which are in continuous rapid, random, straight-line motion. Collisions between the molecules in a gas are elastic (i.e., there is no net loss of kinetic energy on collision), and the average kinetic energy of the molecules in a gas is directly proportional to temperature.

lattice point The center of the specific positions occupied by the atoms, molecules, or ions in a crystalline solid.

law of chemical equilibrium The relationship among the concentrations of reactants and products when a system is at chemical equilibrium; consists of two parts—an equilibrium constant and the mass action expression. At equilibrium, the numerical value of the mass action expression is equal to the equilibrium constant K.

law of conservation of energy In a chemical reaction, energy is always conserved and can be neither created nor destroyed.

law of conservation of matter In a chemical reaction, matter is always conserved and is neither created nor destroyed.

law of constant proportions In every compound, the constituent elements are present in constant proportions by weight.

law of multiple proportions When two elements combine to form more than one compound, the masses of one element that combine with a fixed mass of the other element are in a ratio of small whole numbers.

Le Chatelier's principle When a system at equilibrium is disturbed by a stress, the position of the equilibrium shifts to relieve that stress.

levorotatory isomer An optical isomer that causes rotation of the plane of polarized light in a clockwise direction.

Lewis acid and base A Lewis acid accepts a pair of electrons from another species in a reaction; a Lewis base is the species that donates the pair of electrons. A Lewis acid-base adduct is the compound formed in a Lewis acid-base reaction.

Lewis (electron-dot) structure A representation of molecular structure based on the tendency of an atom to obey the octet rule.

limiting reagent The reactant or reagent that disappears first when two or more reactants undergo reaction; it therefore determines the amount of product formed.

liquids That condensed state of matter characterized by relatively weak bonding between molecules such that they are free to move around one another. Typically, liquids exhibit short-range order as a result of these weak interactions which are not strong enough to give the long-range order characteristic of solids.

London (dispersion) forces Weak intermolecular interactions between instantaneous dipoles on nonpolar molecules.

lone (or **nonbonding**) **pair of electrons** A pair of valence electrons in a molecule that are not involved in bond formation.

macroscopic (or **bulk**) **properties** Those properties of a substance that can be easily measured or observed.

magnetic quantum number (m_l) The quantum number denoting the spatial orientation of an orbital; can have values from $-l$ to $+l$.

manometer A device used to measure pressure.

mass action expression (Q) An expression for a chemical reaction equal to the product of the concentrations of each reaction product raised to a power equal to its coefficient in the balanced equation divided by the product of the concentrations of each reaction reactant raised to a power equal to its coefficient in the balanced equation; at equilibrium, $K = Q$.

melting point The temperature at which a solid changes to a liquid.

Mendeleev's periodic law By observing systematic variations in physical properties such as melting and boiling points, Mendeleev was able to state that the properties of the elements are periodic functions of their atomic weights. With the later revision that arranged the elements in order of increasing atomic number, this became known as the periodic law.

metalloids Elements that have properties intermediate between those of nonmetals and metals and are located directly on either side of the Zintl border in the periodic table.

metals Elements that are lustrous, have high thermal and electrical conductivity, have variable melting and boiling points, are generally ductile and malleable (easily worked), and are located to the left of the Zintl border in the periodic table.

miscible A solution in which the components can be mixed in any proportion.

mixture A combination that consists of elements or compounds (or both) brought together in any proportion and has variable physical properties.

molality (*m*) Concentration expressed as number of moles of solute per kilogram of solvent.

$$m = \frac{\text{no. of moles of solute}}{\text{kg of solvent}}$$

molar ratio A relationship that relates chemically equivalent amounts of any pair of reactants and/or products in the reaction described by a particular equation. Mole ratios are used as conversion factors in setting up calculations involving an equation.

molarity (*M*) Concentration expressed as number of moles of solute per liter of solution.

$$M = \frac{\text{no. of moles of solute}}{\text{1 L of solution}}$$

mole That amount of an element or compound in grams numerically equal to the substance's atomic or formula weight; one mole of a substance contains Avogadro's number of units of that substance.

mole fraction The ratio of the number of moles of a component to the total number of moles present. The mole percent is the mole fraction times 100%.

molecular orbital (MO) An orbital that extends over the entire molecule. When a pair of atomic orbitals (one from each atom involved in a bond) are combined, one bonding MO and one antibonding MO are formed. A bonding MO is one that builds up electron density in the region between nuclei (when filled, it stabilizes the molecule). An antibonding MO has

electron density concentrated outside the region between nuclei (a filled antibonding orbital destabilizes the molecule). A delocalized MO is not confined to two adjacent atoms in a bond but is spread out over several atoms.

molecular solids Solids composed of covalently bonded molecules that are weakly bonded together by intermolecular forces. They have relatively low melting and boiling points and, in general, do not conduct electricity.

molecularity The term that specifies the number of molecules interacting in a single mechanistic step; it is reflected in the overall order of the rate equation.

molecule The smallest characteristic unit of a compound.

Nernst equation An equation that relates standard half-cell potentials and concentrations to half-cell voltage;

$$\mathcal{E} = \mathcal{E}° - \frac{2.30\,RT}{nF}(\log Q)$$
$$= \mathcal{E}° - \frac{0.0591}{n}(\log Q), \text{ at 298 K.}$$

net ionic equation An equation that includes only those species in a solution that enter into the chemical reaction.

neutral solution $[H_3O^+] = [OH^-] = 1.0 \times 10^{-7}$; pH = 7.00 (at 25°C).

neutralization reaction The reaction between an acid and a base mixed in such proportions that the resulting solution has neither acidic nor basic properties.

neutron Neutral particle with a mass of 1.008665 amu (1.6749×10^{-24} g).

noble gas configuration The outer electronic configuration found in the noble gases; a filled *sp* sublevel (ns^2np^6) with the same value of *n*.

noble gases Elements in Group VIIIA (or 18) of the periodic table.

nonelectrolyte A solute that dissolves to give a nonconducting solution.

nonmetals Elements located on the right-hand side of the Zintl border in the periodic table; they tend to be brittle, dull-appearing, and nonconductive.

nonpolar molecule A molecule that does not have a net electric dipole.

normal boiling point The temperature at which the vapor pressure of a liquid is equal to one standard atmosphere (760 torr).

nucleus of the atom The very small, dense, positively charged central portion of the atom that contains protons and neutrons (and therefore most of the mass of the atom).

octet rule A rule that states that many atoms, when entering into bonding, tend to attain a noble gas configuration, ns^2np^6, with a total of eight electrons in the outermost electronic orbitals.

optical activity The rotation of a plane of polarized light as it passes through a chiral substance.

optical isomers Isomers that are non-superimposable mirror images of each other; also known as enantiomers.

osmosis The passage of solvent molecules through a semipermeable membrane into a solution of higher concentration. The osmotic pressure (Π) is the external pressure necessary to prevent osmosis; $\Pi = MRT$.

overall cell potential ($\mathcal{E}$) The sum of the oxidation half-cell potential and the reduction half-cell potential.

overall reaction order The sum of the exponents in the rate law that describes that reaction.

oxidation The process of losing electrons, thereby increasing a species' oxidation number.

oxidation half-cell potential ($\mathcal{E}_{ox}$) The voltage at the anode (oxidation half-cell electrode).

oxidation state (or **oxidation number**) A positive or a negative number assigned to an element according to certain rules.

oxidizing agent The species undergoing reduction. This species consumes the electrons lost by the species undergoing oxidation and decreases in positive charge, going to a lower oxidation state.

oxyacid An oxygen-containing acid.

pair of electrons Two electrons in the same orbital with opposite values of m_s; that is, they have opposite spins.

partial pressure of a gas The pressure that one gas exerts in a mixture of gases; equal to the pressure that the gas would exert if it were the only gas present.

parts per million (ppm) A measure of concentration in dilute solutions. 1 ppm = 1 part solute per 10^6 parts solution, by weight. In aqueous solutions, 1 ppm = 1 mg/L.

Pauli exclusion principle A principle that states that no two electrons in the same atom can have the same four quantum numbers.

percent composition An expression of the elemental composition of a compound. The percent composition of each element (the weight percent) is equal to the weight of that element divided by the total weight present (the formula weight) times 100%.

percent yield The yield obtained in a reaction; expressed as a percentage of the theoretical yield.

period A horizontal row of elements in the periodic table.

pH pH = $-\log$ [H_3O^+]; an expression of the [H_3O^+] of a solution; a measure of acidity.

phase diagram A plot of pressure versus temperature showing the conditions under which a given phase of a substance exists.

phase transition The process by which a substance is transformed from one state (phase) to another state (phase).

photon (or **quantum**) A packet of energy given by $E = h\nu$, where h is Planck's constant and ν is the frequency of the radiation. Light can be considered to be a stream of massless particles, each carrying a quantum of energy proportional to the frequency.

physical equilibrium A dynamic equilibrium that involves a physical transformation.

physical property A property that can be studied without changing the identity of the substance.

pi (π) bond A bond formed by the sideways overlap of a pair of p orbitals. A π molecular orbital is an MO in which electron density lies above and below the internuclear axis.

pipet A calibrated device used to deliver exact amounts of liquid.

Planck's constant (h) The proportionality constant between the frequency and the energy of a photon of light: $h = 6.63 \times 10^{-27}$ erg-s (or 6.63×10^{-34} J-s).

pnictogens (or **pnictides**) Elements in Group VA (or 15) of the periodic table.

pOH pOH = $-\log$ [OH$^-$]; a measure of basicity ([OH$^-$]).

polar bond Bond formed between atoms of different electronegativity.

polar molecule A molecule that has a net electric dipole.

polarizability A measure of the extent to which the electron cloud in an atom or a molecule can be distorted (or polarized).

polyatomic ion An ion composed of two or more atoms. Typically, the bonding between the atoms in a polyatomic ion is covalent, but the charged ion then interacts with oppositely charged ions to form an ionic compound.

potential energy (PE or V) Stored energy or energy resulting from relative position or structure.

potential energy diagram A plot of potential energy versus the reaction coordinate for a chemical reaction.

precipitate The insoluble solid that forms in a solution.

precision of a measurement An expression of the validity of a measurement or how close the measured values are to each other.

pressure Force (mass × acceleration) per unit area.

primary cell An electrochemical cell that can act only as a galvanic cell (it cannot be recharged).

principal quantum number (n) The quantum number that specifies the electron's shell and determines the size of the orbital; it can have only the values 0, 1, 2, 3, . . . , n; roughly corresponds to the integer n in Bohr's equations.

products The species that are produced in a chemical reaction.

proton Positively charged particle with a mass of 1.007276 amu (1.6726 $\times 10^{-24}$ g).

quantum numbers Four constants that designate the energies and locations of electrons in electronic orbitals; see principal, azimuthal, magnetic, and spin quantum numbers.

racemic mixture A mixture that contains equal numbers of the two optical isomers of a substance.

Raoult's law A formal statement of the vapor pressure lowering; the presence of a solute lowers the vapor pressure of the pure solvent in proportion to the solute concentration.

rare earths (or **lanthanides**) The elements cerium through lutetium; formed by filling the $4f$ sublevel.

rate constant (k) The proportionality constant in the rate law for a reaction.

rate-determining step The slow step in a reaction mechanism that determines the rate at which products are formed.

rate law (or **rate equation**) An equation that relates the rate of a reaction to the concentration of the reactants, each raised to the appropriate power. For a reaction involving the species A and B, a general rate law is of the form rate = $k[A]^m[B]^n$, where k is the specific rate constant.

reactants The species undergoing chemical reaction.

reaction coordinate The path followed as reactant molecules approach each other, collide, and make and break chemical bonds, and the product species move apart.

reaction mechanism A detailed model of how a reaction proceeds, including the sequence of steps occurring during the process.

reaction order The exponent in the rate equation for a particular reactant.

reaction rate The rate of a chemical reaction or the rate of decrease in the concentration of a reactant or the increase in the concentration of a product.

reducing agent The species that is undergoing oxidation (it provides the electrons necessary to reduce the other species and, in doing so, loses electrons, increases in positive charge, and goes to a higher oxidation state).

reduction The process of gaining electrons, thereby decreasing a species' oxidation number.

reduction half-cell potential ($\mathcal{E}_{red}$) The voltage at the cathode (reduction half-cell electrode).

representative elements The elements in Groups IA through VIIA (or Groups 1, 2, and 13 through 17) of the periodic table.

resonance Occurs when two or more equivalent Lewis structures can be drawn for the same molecule, each of which is called a resonant structure. The actual structure of a molecule for which resonant structures can be written is known as a resonant hybrid.

retrograde soluble The condition where solubility decreases with increasing temperature.

salt A neutral compound that is neither acidic nor basic; the result of a neutralization reaction in aqueous solution.

salt bridge A tube containing an electrolyte that connects the two half-cells of an electrochemical cell through which ions can move.

saturated solution A solution that has dissolved solute in dynamic equilibrium with undissolved solute and contains the maximum amount of dissolved solute at a particular temperature.

Schrödinger equation A partial differential equation that describes the total energy of electronic systems.

second law of thermodynamics When a spontaneous change occurs, it is accompanied by an increase in the total entropy of the universe: $\Delta S_{universe} = \Delta S_{system} + \Delta S_{surroundings} > 0$

secondary cell An electrochemical cell that can act as a galvanic cell and as an electrolytic cell (it can be recharged).

semiconductor A substance that is a weak conductor of electricity.

semipermeable membrane A membrane through which only certain ions or molecules can pass.

SI (or **International System of Units**) Provides a defined set of seven base units (mass, length, time, temperature, amount of substance, electric current, and luminous intensity) from which all other units can be derived.

sigma (σ) bond A bond formed by head-to-head overlap of two orbitals; is spherically symmetric about the internuclear axis. A σ molecular orbital is an MO in which electron density is spherically symmetric about the internuclear axis.

significant figures Number of meaningful digits in a measured quantity; reflects the precision of a measurement.

single bond A covalent bond in which one pair of electrons is shared.

solubility The degree to which a compound dissolves in a solvent; generally expressed either in moles of solute per liter of solution (molarity) or in grams of solute per 100 g of solution.

solubility product (K_{sp}) An equilibrium constant that is a measure of the solubility of a compound.

solute The minor component in a two-component solution.

solution Another term for a homogeneous mixture. An aqueous solution is one in which water is the solvent.

solvent The major component in a two-component solution.

specific gravity The ratio of the density of a substance to the density of water ($1.000 \, g/cm^3$ at 25°C).

specific heat The amount of heat needed to raise the temperature of 1 g of a substance by 1°C.

spectator ion An ion that is present in a solution but does not participate in the chemical reaction.

spin quantum number (m_s) The quantum number that determines the direction in which the electron seems to be spinning; can have a value of $+\frac{1}{2}$ or $-\frac{1}{2}$.

standard free energy change of formation ($\Delta G_f°$) The change in free energy when one mole of a compound in its standard state is formed from its elements in their standard states.

standard heat of formation ($\Delta H_f°$) The enthalpy change when one mole of a compound in its standard state is formed from its elements in their standard states.

standard hydrogen electrode A half-cell used as a reference electrode; its standard potential is defined as exactly 0.000 V; $2 \, H^+(aq) + 2 \, e^- \rightarrow H_2(g)$.

standard potential ($\mathcal{E}°$) The potential of a galvanic cell under standard conditions, that is, $T = 25°C$, all concentrations are equal to $1.00 \, M$, and the partial pressure of any gas = 1.00 atm. The standard oxidation potential ($\mathcal{E}°_{ox}$) is the potential for a given anodic half-reaction relative to the hydrogen electrode under standard conditions; the standard reduction potential ($\mathcal{E}°_{red}$) is the potential for a given cathodic half-reaction.

standard state 1 atm pressure and 25°C.

standard temperature and pressure (STP) The standard temperature and pressure for a gas: 273 K (0°C) and 1.00 atm.

state function A quantity whose value depends only on the current state of a system; a change in its value depends only on the initial and final states and is independent of path.

strong acid An acid that is essentially completely dissociated in solution; a strong electrolyte.

strong electrolyte A solute that is essentially completely ionized in solution, giving a solution that is a good conductor of electricity.

structural isomers Isomers that differ in the sequence in which the atoms are bonded together.

sublimation The transition from a solid directly to a gas.

substitutional solid solution A solid solution in which an atom or molecule of the solute replaces an atom or molecule of the solvent in its lattice.

supercooled liquid A liquid at a temperature below its normal freezing point.

supercritical fluid A substance existing at a temperature above its critical temperature.

supersaturated solution An unstable solution that contains more solute than it would in the equilibrium state.

surface tension Phenomenon that results from attractive forces trying to pull molecules from the surface of a liquid into the bulk of the liquid.

surroundings In thermodynamics, everything in the universe outside the system being studied.

system In thermodynamics, the part of the universe that is being studied. A closed system is a system that allows the exchange of energy, but not mass, with its surroundings.

table of relative acid-base strengths A listing of acids in order of the ease with which they transfer a proton to the solvent water (i.e., their strength).

temperature The degree of hotness or coldness of matter, such as a body or a fluid.

theoretical yield The maximum amount of product that can be produced in a given reaction under ideal conditions.

third law of thermodynamics The entropy of a pure crystalline substance at absolute zero is zero; $S = 0$ at $0\,K$.

titration The reaction of an amount of a substance with an exact amount of a second substance until the complete reaction occurs.

transition metals (or **transitional elements**) The elements in Groups IB through VIIIB (or Groups 3 through 12) of the periodic table; formed by filling the d sublevels.

triple bond A covalent bond in which three pairs of electrons are shared.

triple point That value of T and P at which there is equilibrium among solid, liquid, and gas phases of a substance.

Tyndall effect The scattering of light by a colloidal suspension because the size of the suspended particle is on the order of magnitude of visible light.

unit cell The smallest part of a lattice that can be repeated over and over in all directions to give the entire crystal lattice. The edges of the unit cell are defined as a, b, and c, and the angles between them as α, β and γ (α is the angle between edges b and c, and so on). The seven simple unit cells are cubic, tetragonal, orthorhombic, hexagonal, rhombohedral, monoclinic, and triclinic.

unsaturated hydrocarbon A hydrocarbon that has double or triple bonds; additional hydrogen atoms can be added to the carbon atoms.

unsaturated solution A solution that contains less than the amount of solute in a saturated solution at a particular temperature and is therefore capable of dissolving additional solute.

valence bond A bond formed by sharing a pair of electrons between two overlapping atomic or hybrid orbitals.

valence electrons Electrons in the outer electronic configuration of an atom (outside the noble gas configuration).

valence shell electron pair repulsion (VSEPR) theory A theory for the prediction of molecular structure that is based on the tendency of electron pairs in the valence shell to be as far apart as possible.

Van der Waals equation A modification of the ideal gas law in which corrections are made for the volume of molecules and the presence of intermolecular forces.

vapor pressure The pressure exerted by the vapor over a liquid.

vapor pressure lowering The decrease in the vapor pressure of a solution below the vapor pressure of pure solvent.

visible light That part of the electromagnetic spectrum to which the human eye is sensitive, ranging from about 4×10^{-5} cm (4000 Å) to 7×10^{-5} cm (7000 Å).

volume percent The volume of solute divided by the volume of solution times 100%.

$$\text{vol }\% = \left(\frac{\text{vol of solute}}{\text{vol of solution}} \right) \times 100\%$$

wave function (ψ) A mathematical expression that is a solution to the Schrödinger equation and that describes the behavior of an electron in an atom. The square of the wave function (ψ^2) is a probability function

that gives the probability of locating an electron in a given volume of space.

wavelength (λ, lambda) The distance between the crests of a wave.

wave mechanical picture of the atom A model for the atom that considers electrons to have a wavelike character.

wave-particle duality The concept by which light can be described equally well and simultaneously as if it has a wavelike and particulate character.

weak acid An acid that is only partially dissociated in solution; a weak electrolyte.

weak electrolyte A solute that is only slightly ionized in solution, giving a solution that is a weak electrical conductor.

weight The weight of an object is equal to the mass of that object times the force of gravity (g).

work Results from the transfer of energy. To do work, energy must be expended.

X-ray diffraction The scattering of X rays by the array of species comprising a crystalline solid that form planes of lattice points. The distance between these planes (d) is in the range of 2–4 Å, which is in the X-ray region of the spectrum.

zero-order reaction A reaction whose rate is independent of concentrations.

Zintl border The zigzag line that separates the metallic elements from the nonmetallic elements in the periodic table.

INDEX